Anglo-French Physical Acoustics Conference (AFPAC 2024)

Journal of Physics: Conference Series Volume 2904

Loch Lomond, United Kingdom
17-19 January 2024

ISBN: 979-8-3313-1037-0
ISSN: 1742-6588

Printed from e-media with permission by:

Curran Associates, Inc.
57 Morehouse Lane
Red Hook, NY 12571

CURRAN ASSOCIATES INC. proceedings.com

Some format issues inherent in the e-media version may also appear in this print version.

Institute of Physics
Dirac House, Temple Back
Bristol BS1 6BE UK

Phone: 44 1 17 929 7481
Fax: 44 1 17 920 0979

techtracking@iop.org

Additional copies of this publication are available from:

Curran Associates, Inc.
57 Morehouse Lane
Red Hook, NY 12571 USA
Phone: 845-758-0400
Fax: 845-758-2633
Email: curran@proceedings.com
Web: www.proceedings.com

TABLE OF CONTENTS

Journal of Physics: Conference Series **2904** (2024) 011001 doi:10.1088/1742-6596/2904/1/011001

Anglo-French Physical Acoustics Conference (AFPAC) 2024
17 to 19 January 2024
Lodge on Loch Lomond Hotel, Luss, Scotland

The Anglo-French Physical Acoustics Conference (AFPAC) is organised by the IOP Physical Acoustics Group, IOP Medical Physics Group, Le Groupe d'Acoustique Physique and the Sous-marine et UltraSonore (GAPSUS) of the Société Française d'Acoustique (SFA). AFPAC provides a forum in which the current research activity in physical acoustics and ultrasonics is reviewed.

The scope for this conference is intentionally wide to encourage cross-fertilisation of ideas across different application areas in the field of physical acoustics. Topics include, but are not limited to: theoretical developments (all methods); experimental measurements; transduction; linear or nonlinear elastic or acoustic (bulk, surface, guided) wave radiation, propagation, scattering and reception; application to material characterisation, biomedical ultrasound, imaging (nondestructive testing, medical, underwater, seismology), SAW in electronic devices etc.

This year AFPAC is additionally organised with the IOP Medical Physics group and includes sessions focussed on Biomedical ultrasound, spanning both medical imaging and therapy. We are also pleased to announce two invited speakers from the field of biomedical ultrasound: Professor Gail ter Haar from the Institute of Cancer Research and Professor Nader Saffari from University College London.

Committee:

- **Theodosia Stratoudaki**, University of Strathclyde, UK (IOP - Physical Acoustic Group)
- **Emma Harris**, Institute of Cancer Research (ICR), UK (IOP - Physical Acoustic Group)
- **Nader Saffari**, University College London, UK (IOP - Physical Acoustic Group)
- **Peter Huthwaite**, Imperial College London, UK (IOP - Physical Acoustic Group)
- **Paul Prentice**, University of Glasgow, UK (IOP - Physical Acoustic Group)
- **Helen Mulvana**, University of Strathclyde, UK (IOP - Physical Acoustic Group)
- **Samuel Raetz**, Le Mans University, France (SFA - GAPSUS)
- **Alain Lhémery**, CEA Paris-Saclay, France (SFA - GAPSUS)
  Dr Bahadar Bhatia, Sandwell and West Birmingham NHS Foundation Trust, UK (IOP – Medical Physics Group)
- **Dr Caryl Richards**, University of Leicester, UK (IOP – Medical Physics Group)

Journal of Physics: Conference Series **2904** (2024) 011002 doi:10.1088/1742-6596/2904/1/011002

Peer Review Statement

All papers published in this volume have been reviewed through processes administered by the Editors. Reviews were conducted by expert referees to the professional and scientific standards expected of a proceedings journal published by IOP Publishing.

- **Type of peer review:** Single Anonymous
- **Conference submission management system:** Morressier
- **Number of submissions received:** 9
- **Number of submissions sent for review:** 9
- **Number of submissions accepted:** 8
- **Acceptance Rate (Submissions Accepted / Submissions Received × 100):** 89%
- **Average number of reviews per paper:** 4
- **Total number of reviewers involved:** 4
- **Contact person for queries:**
  Name: Vivien Thomas
  Email: vivien.thomas@iop.org
  Affiliation: Institute of Physics

Journal of Physics: Conference Series **2904** (2024) 012001 doi:10.1088/1742-6596/2904/1/012001

Inverse problem-solving method applied to acoustic characterisation of a thin film using Debye series model

Pooyan MANOOCHEHRNIA, Pierre MARECHAL, Damien LEDUC, Mounsif ECH-CHERIF EL-KETTANI

LOMC, UMR 6294 CNRS, University of Le Havre Normandy, 75 Rue de Bellot, 76600 Le Havre, France

pooyan.manoochehrnia@gmail.com

Abstract

In this paper, it is proposed to solve an inverse problem with the perspective of determining the ultrasonic properties (thickness, longitudinal wave velocity, density and attenuation coefficient) of a thin epoxy resin film deposited on a glass substrate with well-known properties. The question focuses on finding acoustic parameters of the structure and the transducer thanks to a numerical implementation of a physical model matching the experimental ultrasonic response. The Debye Series Model (DSM) has been used to model the structure. A step by step procedure is implemented where at each step one parameter is estimated by minimizing the MSE (Mean Square Error) between simulated model and experimental signal. Estimated parameters are checked to be physically viable. At beginning an initial set of parameters are announced together with related waveforms and at the end the final adjusted parameters are presented beside finally obtained waveforms.

1. Introduction

In recent years, the analysis of the non-uniqueness and uncertainty of solutions to inverse problems has played an increasing role in physics modelling. Linear problem-solving analysis has been well established for many years, while non-linear problem analysis methods have been developed more recently. Even if best-fit methods give solutions, the analysis of uncertainty and uniqueness is very important for the evaluation of scientific conclusions based on inverse calculations. The inverse problem resolution can be carried out methodically by an a priori step-by-step procedure, leading to a unique set of parameters. As a general description of the problem, a modelling method is used as a description of the physical problem, on the basis of a given set of input parameters. In ultrasound modelling of multi-layered structures, the studied structure could be either modelled in oblique incidence using the stiffness matrix model (SMM) as mentioned in [1], [2], [3], [4] or in normal incidence using Debye series model (DSM) as mentioned in [5], [6], [7]. In this paper, the structure has been modelled using DSM which is unconditionally stable and for any type of configuration as first described in [5], [8], [9], [10].

The reflection spectrum of the multilayer structure calculated via DSM is combined to the impulse response spectrum of the transducer [11], [12], [13], by a multiplication in the frequency domain.

In summary, DSM involves calculation of the global reflection coefficient as a function of frequency, $R_g(f)$ (1), of the multilayer structure based on global transmission coefficient, $T_g(f)$ (2), iterative calculation of local reflection, $R_{p,q}(f)$, and transmission, $T_{p,q}(f)$, coefficients of each layer, (3) and (4). Equations (1) to (4), are calculated using equations (5) to (7). Once the global reflection coefficient, R_g, has been calculated, it will be multiplied by spectrum of incident signal which is usually in form of a gaussian function. Finally, the inverse Fourier transform of the later function is calculated which in turn results in time-based echo signal addressing the structure response to ultrasonic acoustic solicitation.

$$R_g = R_{12} + \frac{T_{12}\rho_{23}T_{21}}{1 - R_{21}\rho_{23}} \tag{1}$$

$$T_g = T_{12}\prod_{n=2}^{n=N} \frac{T_{n,n+1}}{1 - R_{n,n+1}\rho_{n,n-1}} \tag{2}$$

$$\begin{cases} R_{n,n+1} = \dfrac{Z_{n+1} - Z_n}{Z_{n+1} + Z_n} e^{2j\varphi_n} \\ T_{n,n+1} = \dfrac{2Z_{n+1}}{Z_{n+1} + Z_n} e^{-j(\varphi_{n+1} - \varphi_n)} \end{cases} \tag{3}$$

$$\begin{cases} R_{n+1,n} = \dfrac{Z_n - Z_{n+1}}{Z_n + Z_{n+1}} e^{-2j\varphi_{n+1}} \\ T_{n+1,n} = \dfrac{2Z_n}{Z_{n+1} + Z_n} e^{-j(\varphi_{n+1} - \varphi_n)} \end{cases} \tag{4}$$

$$\begin{cases} \varphi_n = k_n Th_n = \dfrac{\omega Th_n}{C_n} \\ \varphi_{n+1} = k_{n+1} Th_n = \dfrac{\omega Th_n}{C_{n+1}} \end{cases} \tag{5}$$

$$\begin{cases} \rho_{n+1,n} = R_{n+1,n} + \dfrac{T_{n+1,n}\rho_{n,n-1}T_{n,n\;1+}}{1 - R_{n,n+1}\rho_{n,n-1}} \quad \text{for} \quad 2 \le n \le N \\ \rho_{21} = R_{21} \end{cases} \tag{6}$$

$$\begin{cases} \rho_{n,n+1} = R_{n,n+1} + \dfrac{T_{n,n+1}\rho_{n+1,n+2}T_{n+1,n}}{1 - R_{n+1,n}\rho_{n+1,n+2}} \quad \text{for} \quad 2 \le n \le N \\ \rho_{N+1,N+2} = R_{N+1,N+2} \end{cases} \tag{7}$$

Two sets of parameters are dealt with in this paper, namely transducer envelope parameters and structure acoustic parameters. The parameters are identified and classified in such a way that their effect on the final simulation result is isolated. The mean square error (MSE) between simulated and experimental signals is taken as an index of similitude between two signals, before passing from one set of parameters to the next [14], [15]. Above mentioned methods lead to estimation of acoustic parameters like thickness, speed, density and other elastic constants of various layers. Although in this study scanning acoustic microscopy (SAM)has been used in reflection mode, its usage has been concentrated only on punctual measurements (A-Scan).

2. Experiment

An 85 μm thick layer of epoxy resin has been deposited on a glass plate of 1.66 mm thick to form a two-layered structure. The thin film of epoxy-resin is deposited on the glass plate using a spin-coater with a pre-set turning speed of 2000 rpm and duration of 20 s [21], [22], [23]. The thickness of deposit is measured using a precise roughness meter device (Mitutoyo Surftest SV-2100). Surface profile obtained from roughness meter device illustrates that the thickness profile of the deposit is not uniform. Thus, a relatively uniform part of deposit has been marked to be characterised using SAM. This thickness value is then used in procedure of calculations. In addition, on the basis of several sources, different values for density of epoxy resin are reported and vary within a very large range. Nevertheless, considering the specific combination of resin and epoxy hardener that is used in this experiment, the value of the density of the actual epoxy resin was estimated at 980 ± 2% [kg/m3]. Each layer of the structure is to be defined by its geometrical and acoustical parameters including the thickness Th, the longitudinal acoustic wave velocity C, the density ρ and the acoustic attenuation coefficient α. The structure is characterised using the SAM equipped with a plane wave transducer with a central frequency of 50 MHz working as emitter and receiver. The structure is immersed in water together with the transducer which measures the acoustic echo reflected at normal incidence from the structure. A dedicated software has been developed by SAM manufacturer (WinSAM v8) for the driving of the translations along the (x, y, z) axes with the declared precision of 1 μm, as well as for the capture of the ultrasonic echoes. The typical configuration setup is composed of a plane wave transducer with central frequency of 50 [MHz] driven with an integrated electric impulse of 5 [ns] duration, has been positioned within a distance equivalent to ToF (Time of Flight) of 8.9 [μm] from the structure under test. An oscilloscope with sampling frequency of 25 [GHz] and quantification of 8 [Bits] is used for this measurement.

3. Method

3.1. Model Calculation process

The electroacoustic response of the studied structure is simulated via an efficient numerical model based on the Debye series decomposition and named as DSM [6, 7, 8]. This DSM tool calculates the reflection coefficient of the whole structure on the basis of series of the local transmission and reflection coefficients at each interface between the layers, and as function of the frequency [6], [12]. The local reflection R_{mn} and transmission T_{mn} coefficients are calculated based on expressions defined in the DSM formalism. Those expression are based on the properties of the multi-layer structure, *i.e.* acoustic impedance Z_n, and the Fresnel phase parameters. Finally, the total reflection coefficient R_g is calculated. Then the resulting reflection spectrum R_g is then multiplied by a gaussian envelope G_e representing the electroacoustic response of the transducer as a symmetric envelope. The time-domain electroacoustic response $s(t)$ is obtained as a result of the numerically implemented inverse Fourier transform.

3.2. Inverse problem solving by parameter matching process

The DSM has been implemented using MATLAB and it results in the electroacoustic response of a given transducer (G_e) on a given multilayer structure (R_g and T_g). The resulting modelled signal $s(t)$ is then compared to the experimental one. The global electroacoustic response includes each reflected echo in the bilayer. This imposes simultaneous characterisation of acoustic parameters of both substrate (variables indexed s) and deposit (variables indexed d). Nevertheless, as we are only interested by the properties of the thin layer (those of the glass substrate are previously well identified), only the first package of echoes due to multiple reflections in the thin film are considered by the process of matching between the model and the experiment.

The first step of the inverse problem concerns the transducer's response, which is modelled by the central frequency f_0 and the relative bandwidth $\Delta f_{6,r}$, corresponding (in a first approach in the time domain) to the pseudo-period and the echo duration. In addition, the phase shift φ_s of the transducer's response is updated to match to the local minima and maxima of the echoes.

The second step consists in the syncing of time of flight ToF inside the deposit layer ($ToF_d = Th_d / C_d$), between experimental signal and DSM resulting in, in one hand inserting a reference point in signals for further adjustments of two signals and on the other hand resulting in determination of parameter pair of C_d and Th_d.

The third step is the determination of the acoustic impedance of the deposited thin film $Z_d = \rho_d C_d$ that allows the better matching of the modelled and experimental signals amplitudes. In the fourth step, the attenuation in the thin film is adjusted so that the mismatch is minimum. The amplitudes of the subsequent echoes are directly related to the attenuation of the deposit layer α_d, which can be finally adjusted.

The objective is to start from the above defined configuration with the input set of parameters $\{\rho_d, C_d, \delta_d, Th_d\}$ for the deposit and modify it in such a way that the DSM matches to the experimental signal. To do this, a step-by-step method have been set up to provide a robust and reproducible result. Knowing with a good confidence some of the parameters (substrate properties), the described step-by-step preliminary approach ensures the unicity of the initial set of physical parameters. This set of initial values of properties are summarized inTable 1.

Table 1: Initial parameters: Fixed properties of the substrate and initial values of the deposit.

Parameter	ρ [kg/m^3]	C [m/s]	δ [%]	Th [µm]
Substrate	2500	5800	0	1.6
Deposit	1300	2500	1.5×10^{-3}	85

A zoom on the first package of echoes which illustrates the signal resulting from the simulation of the studied structure with above parameters, and experimental signal is shown in Fig. *1*.

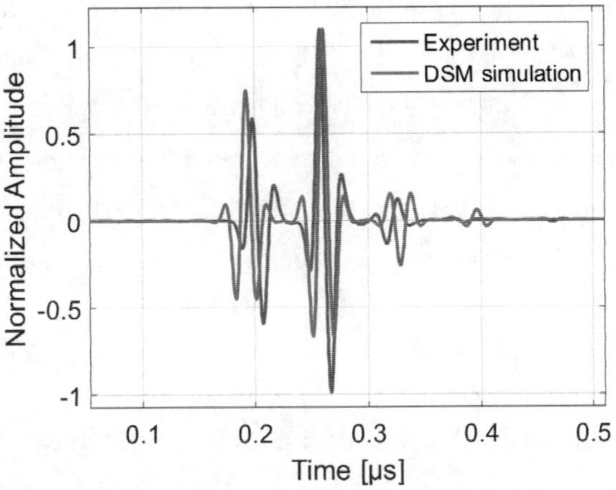

Fig. 1: DSM modelled and SAM experimental comparison between the first package of echoes for initial state.

During the procedure following steps are followed:

3.2.1. Step 1: ToF_d

The time-of-flight corresponding to above simulated parameters is $ToF_d = 2Th_d / C_d = 68$ ns, whereas experimentally, this ToF_d is clearly lower. A time delay estimation leads to reduce this ToF_d by 15% of its initial value. This reduction of 15% could be done either by reducing the thickness Th_d or by increasing C_d or any linear combination of both. Therefore, the thickness at this step is multiplied by 0.85. At this step, the reduction of –15% in TOF is fixed and the algorithm is followed to the next step.

3.2.2. Step 2: φ

From the result of previous step, it is observed that even if the envelopes of the echoes are in phase, the phase shift of the 1st roundtrip echo of the DSM signal is unlike the experiment. This could be corrected throughout the transducer's impulse response G_e by introducing phase shift φ to the gaussian envelope used in DSM. Indeed, the desired phase shift inducing the required phase shift should be between 0 and π or between π and 2π. Therefore, the value of $3\pi / 2$ is tested which results in better match between two signals. With an additional refinement, the adjusted phase shift would be equivalent to $3\pi / 2 \times 1.08$, which results in an excellent match between two signals. Therefore, the phase shift is fixed to above value and the algorithm is followed to the next step.

3.2.3. Step 3: C_d

It should be noted that the notion of acoustic impedance plays an important role here. It is possible to keep the ToF_d fixed (as set in step 1) but to change the amplitude of signal in order to get a better match between simulated model and experimental signal. This can be done by multiplying both thickness and celerity to the same value. For example, thickness could be multiplied to 0.85 (previously set) and then again multiplied to 0.9. In order to keep ToF_d fixed, at the same time, C_d should be multiplied by 0.9. This will result in a better match between experimental and simulated signals.

3.2.4. Step 4: Z_d

From the above statement, it could be deduced that in order to get a better match between the amplitudes, the acoustic impedance should be changed. This means that the change in ToF_d affects the acoustic impedance Z_d via the velocity C_d. Therefore, it should be compensated by the density ρ_d to reach to the desired overall change in Z_d and ToF_d at the same time. For example, one compromise would be to evenly distribute the 15% reduction in ToF_d, that has been already fixed in step 1, as a linear combination between the velocity C_d and the thickness

Th_d, e.g. –8% on Th_d and +7% on C_d. Above discussion leads us obligatory to a +7% increase in Z_d that is compensated by –10% change in ρ_d resulting in total –3% change in Z_d. Although this could result in good numerical match between curves, it may not result in satisfactory estimated physical values for the materials. Therefore, we need some fine adjustments as follows.

3.2.5. Step 5: Refinement

As a refinement of step 4, the set of parameters $\{\rho_d, C_d, \delta_d, Th_d\}$ for the deposit improves the fit in the numerical sense, but is not satisfying physically. A fine-tuned adjustment around the previous values is necessary by modifying the results as in Table 2. It should be noted that the key for adjustment of these parameters is the known physical value of density $\rho = 980 \pm 2\%$ [kg/m3], as previously mentioned in section 2. Observing the resulting adjustment, a bigger mismatch between amplitudes is observed. This mismatch is the price to be paid for physically acceptable results. Nevertheless, the mismatch could be also due to inappropriate attenuation factor.

Table 2: Parameters values used in Step 5.

Parameter	ρ [kg/m^3]	C [m/s]	δ [%]	Th [μm]
Step 5 values	1300×0.75	2500×1.07	1.5×10^{-3}	85×0.93

3.2.6. Step 6: δ_d

In this step, the loss coefficient δ_d in the deposit is adjusted to fit the third echo, so that its order of magnitude is between 1% and 10%. As a result, the loss coefficient is found at $\delta_d = 1.5\%$. It should be noted that the attenuation value α_d is directly related to the loss coefficient δ_d that is the relative imaginary part of the velocity C_d. It is usually written as $C_d = C_{0d}.(1 + j\delta_d)$. This velocity loss coefficient δ_d can be directly related to the attenuation $\alpha_d = \alpha_{0d}.f$, implicitly considered as a frequency dependent coefficient. As a result, the complex wavenumber in the deposit can be written as: $k = \omega / C_d + i.\alpha_d$ and the slope of the attenuation coefficient has been identified to be $\alpha_{0d} = 3.5\times10^{-5}$ Np.m^{-1}.Hz^{-1}.

3.2.7. Step 7: φ refinement

The previous changes result in a time shift between the 1st peaks of 0.5 ns. Therefore, an additional phase-shift correction is applied to the gaussian envelope, therefore multiplying the

Journal of Physics: Conference Series **2904** (2024) 012001 doi:10.1088/1742-6596/2904/1/012001

phase-shift by a further 1.04 coefficient resulting in $\varphi_d = 3\pi/2\times1.08\times1.04$. As illustrated in Fig. 2, it gives a much better match than all previous results.

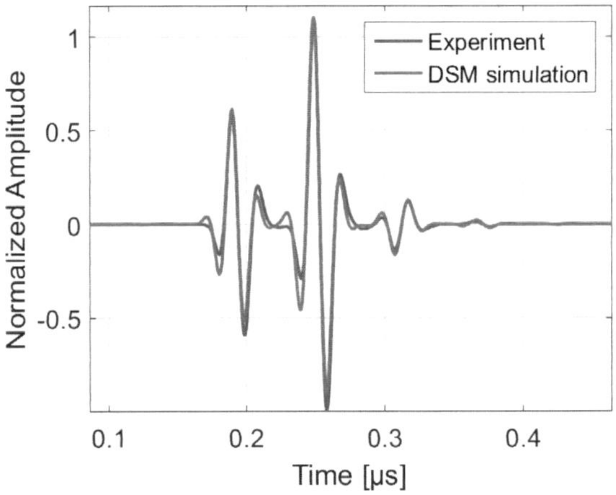

Fig. 2: DSM modelled and SAM experimental signals using parameters in Step 7.

4. Results & Discussion

In summary, the developed algorithm is divided in following two parts:

A. Initialization: in this section the maximum peaks are synchronized and then an initial set of 11 parameters is set. Then the algorithm is focused on the 1st package of echoes bypassing the subsequent return echoes of substrate momentarily at this stage.

B. Calculating the change percentage needed to apply on each of the deposit parameters $\{\rho_d, C_d, \delta_d, Th_d\}$ and on gaussian frequency envelope representing the transducer bandwidth $\{f_0, \varphi, \Delta f_6\}$. Continuing to update above parameters as long as the mean square error (MSE) has not been decreased by a specified percentage.

The final matched parameters are summarized in Table 3.

Table 3: Finally matched set of parameters.

Parameter	φ [Rad]	ρ [kg/m^3]	C [m/s]	δ [%]	Th [μm]
Final value	$3\pi/2\times1.08\times1.04$	1300×0.75	2500×1.07	1.5	85×0.93

As a summary, film thickness and other viscoelastic parameters of a deposited thin film on a known substrate could be deduced by the step-by-step method presented in this paper. These parameters are occasionally inferred jointly due to their joint physical constraints.

5. Conclusion

A very good agreement between form of signals is achieved, since the time shift between echoes has been corrected, and the amplitudes have been matched. This results in a very good match between DSM simulation results and SAM experimental signals which presents a promising methodology for solving the inverse-problem of characterisation of thin layer using SAM.
As a perspective similar analytical studies are being explored with higher order reflections considered. Additionally, these data could be exploited as additional information for the viscoelastic behaviour, i.e. frequency dependence of the wave velocity and attenuation.

6. References

[1] B. Hosten, "Bulk heterogeneous plane waves propagation through viscoelastic plates and stratified media with large values of frequency domain," *Ultrasonics*, vol. 29, no. 6, pp. 445–450, Nov. 1991, doi: 10.1016/0041-624X(91)90074-I.
[2] B. Hosten and M. Castaings, "Surface impedance matrices to model the propagation in multilayered media," *Ultrasonics*, vol. 41, no. 7, pp. 501–507, Sep. 2003, doi: 10.1016/S0041-624X(03)00167-7.
[3] S. I. Rokhlin and L. Wang, "Stable recursive algorithm for elastic wave propagation in layered anisotropic media: Stiffness matrix method," *The Journal of the Acoustical Society of America*, vol. 112, no. 3, pp. 822–834, Sep. 2002, doi: 10.1121/1.1497365.
[4] L. Wang and S. I. Rokhlin, "Stable reformulation of transfer matrix method for wave propagation in layered anisotropic media," *Ultrasonics*, vol. 39, no. 6, pp. 413–424, Oct. 2001, doi: 10.1016/S0041-624X(01)00082-8.
[5] J.M. Conoir and P. CEDOCAR, "Réflexion et transmission par une plaque fluide," in *La diffusion acoustique*, Gespa., vol. Chap. 5, 1987.
[6] P. Marechal *et al.*, "Modeling of a high frequency ultrasonic transducer using periodic structures," *Ultrasonics*, vol. 48, no. 2, Art. no. 2, Apr. 2008, doi: 10.1016/j.ultras.2007.11.007.
[7] P. Marechal, O. Lenoir, A. Khaled, M. Ech Cherif El Kettani, and D. Chenouni, "Viscoelasticity effect on a periodic plane medium immersed in water," *Acta Acustica united with Acustica*, pp. 1036–1043, 2014.
[8] J. Conoir, "Séries de debye et interférences dans une plaque élastique," *Journal de Physique IV Proceedings*, p. pp.C1-695-C1-704, 1992. [Online]. Available: https://doi.org/10.1051/jp4:19921151
[9] J.M. Conoir, "Acoustic scattering from cylinders, Ph.D Thesis (in French).," Paris VI University, 1987.
[10] J. N. Munday, C. B. Bennett, and W. M. Robertson, "Band gaps and defect modes in periodically structured waveguides," *The Journal of the Acoustical Society of America*, vol. 112, no. 4, pp. 1353–1358, Oct. 2002, doi: 10.1121/1.1497625.
[11] N. Ghodhbani, P. Marechal, and H. Duflo, "Curing and Post-curing Viscoelastic Monitoring of an Epoxy Resin," *Physics Procedia*, vol. 70, pp. 106–109, 2015, doi: 10.1016/j.phpro.2015.08.053.

[12] A. Khaled, P. Maréchal, O. Lenoir, M. Ech-Cherif El-Kettani, and D. Chenouni, "Study of the resonances of periodic plane media immersed in water: Theory and experiment," *Ultrasonics*, vol. 53, no. 3, pp. 642–647, Mar. 2013, doi: 10.1016/j.ultras.2012.11.011.
[13] E. B. Ndiaye, P. Maréchal, and H. Duflo, "Adhesion characterization and defect sizing of sandwich honeycomb composites," *Ultrasonics*, vol. 62, pp. 103–111, Sep. 2015, doi: 10.1016/j.ultras.2015.05.007.
[14] A. E. Mouhtadi, J. Duclos, and H. Duflo, "Ultrasonic modeling of a viscoelastic homogeneous plate," 2012.
[15] T. E. Gómez Álvarez-Arenas, "Simultaneous determination of the ultrasound velocity and the thickness of solid plates from the analysis of thickness resonances using air-coupled ultrasound," *Ultrasonics*, vol. 50, no. 2, pp. 104–109, Feb. 2010, doi: 10.1016/j.ultras.2009.09.009.
[16] Z. E. A. Fellah *et al.*, "Application of the Biot model to ultrasound in bone: Inverse problem," *IEEE Trans. Ultrason., Ferroelect., Freq. Contr.*, vol. 55, no. 7, pp. 1516–1523, Jul. 2008, doi: 10.1109/TUFFC.2008.827.
[17] N. Sebaa *et al.*, "Ultrasonic characterization of human cancellous bone using the Biot theory: Inverse problem," *The Journal of the Acoustical Society of America*, vol. 120, no. 4, pp. 1816–1824, Oct. 2006, doi: 10.1121/1.2335420.
[18] O. Behrend, A. Kulik, and G. Gremaud, "Characterization of thin films using numerical inversion of the generalized Lamb wave dispersion relation," *Applied Physics Letters*, vol. 62, no. 22, pp. 2787–2789, May 1993, doi: 10.1063/1.109209.
[19] M. R. Karim, A. K. Mal, and Y. Bar-Cohen, "Inversion of leaky Lamb wave data by simplex algorithm," *The Journal of the Acoustical Society of America*, vol. 88, no. 1, pp. 482–491, Jul. 1990, doi: 10.1121/1.399927.
[20] P.-C. Xu, A. K. Mal, and Y. Bar-Cohen, "Inversion of leaky Lamb wave data to determine cohesive properties of bonds," *International Journal of Engineering Science*, vol. 28, no. 4, pp. 331–346, Jan. 1990, doi: 10.1016/0020-7225(90)90106-S.
[21] U. G. Lee, W.-B. Kim, D. H. Han, and H. S. Chung, "A Modified Equation for Thickness of the Film Fabricated by Spin Coating," *Symmetry*, vol. 11, no. 9, p. 1183, Sep. 2019, doi: 10.3390/sym11091183.
[22] C.-C. Chang, C.-L. Pai, W.-C. Chen, and S. A. Jenekhe, "Spin coating of conjugated polymers for electronic and optoelectronic applications," *Thin Solid Films*, vol. 479, no. 1–2, pp. 254–260, May 2005, doi: 10.1016/j.tsf.2004.12.013.
[23] C. J. Lawrence and W. Zhou, "Spin coating of non-Newtonian fluids," *Journal of Non-Newtonian Fluid Mechanics*, vol. 39, no. 2, pp. 137–187, Jan. 1991, doi: 10.1016/0377-0257(91)80010-H.

High-resolution ultrasonic characterization of an adhesive film in an aeronautical assembly

Youness EZZIANI[1], Mounsif ECH-CHERIF EL-KETTANI[1], Damien LEDUC[1], Pierre MARECHAL[1]
Mathieu DUCOUSSO[2], Nicolas CUVILLIER[2]

1. LOMC, UMR 6294 CNRS, University of Le Havre Normandy, 75 Rue de Bellot, 76600 Le Havre, France
2. Safran Tech, Rue des Jeunes Bois, Chateaufort, 78772 Magny-les-Hameaux, France.

youness.ezziani@univ-lehavre.fr

Abstract

In recent years, the increasing use of composite materials in aeronautical structures to reduce weight has led to a growing interest in non-destructive testing of adhesive joints. Such inspections aim to detect defects such as porosity or delamination and to identify detachments between the adhesive joint and the substrate. This study focuses on determining the properties of an adhesive film in aerospace assemblies involving titanium/composite materials. Specifically, it aims to precisely measure the thickness of the adhesive film and assess the adhesion quality using high-frequency, high-resolution non-destructive ultrasonic methods. The challenge lies in accurately determine the thickness and mechanical properties of the adhesive joint due to the highly contrasting impedance discontinuities in this type of structure: strong between the titanium and the adhesive, and weak between the adhesive and the composite.
Keywords: Non-destructive evaluation; ultrasonic bond testing; scanning acoustic microscopy; adhesive film characterisation; multilayer structures; titanium and composite materials.

1. Introduction

Reducing the weight of aeronautical structures is critical to reducing aircraft fuel consumption and emissions. Adhesive bonding, which offers advantages over conventional methods such as welding or riveting, addresses this need for weight reduction. Consequently, non-destructive evaluation (NDE) [1][2] is required to assess the adhesion quality. Various non-destructive methods are used for this purpose, including the identification of imperfections such as porosity and delamination [3]. This study focuses on a sample representing the Leap engine fan blade, provided by Safran, made of a titanium alloy TA6V (Ti) bonded to a thick 3D woven composite (Comp) using a thin layer of epoxy resin AF191K (Epo) around hundred micrometers. As a result, the studied structure is a trilayer stack referred to as (Ti/Epo/Comp). Due to the composite's thickness and properties, it is considered a semi-infinite medium in the frequency domain of investigation, around tens of MHz. The axial resolution of the chosen transducer must be adequately matched to the thickness of the adhesive layer for effective echo separation, typically achieved by a sufficient thickness-to-wavelength ratio. Therefore, for thin adhesive layers, a transducer with a centre frequency above several tens of MHz is required. Until now, it has been a challenge to obtain a robust information on the level of adhesion, which is strongly dependent on the adhesive thickness and the acoustic impedance ratios. The difficulty in obtaining an ultrasonic evaluation with a favourable signal-to-noise ratio is due to the acoustical impedance contrast. This contrast is high between TA6V and the epoxy resin, and low between the epoxy resin and the composite, making it particularly difficult to detect the reflected echo at the epoxy/composite interface. This evaluation is carried out using a non-destructive evaluation with an ultrasonic method with high frequency, utilizing the scanning acoustic microscope (SAM) PVA TEPLA 301 (pulse-echo method). In Section 2, the samples studied are described. In Section 3, an experimental study is carried out to evaluate the mechanical properties of the epoxy film and the quality of its adhesion.

2. Description of the experimental studied samples

The samples under investigation are representative of the Leap engine fan blade and are provided by Safran. They are composed of three materials: a titanium alloy TA6V (Ti) bonded to a thick 3D woven composite (Comp) using an epoxy resin AF191K (Epo). The related epoxy bonding process involves cleaning the interfaces with ethanol. Next, strips of a given material and thickness are placed on the edges of the surfaces between the two materials to control the thickness of the epoxy layer and prevent the adhesive from leaking during polymerisation [1]. Finally, polymerisation is carried out in a pressure-controlled oven, where four calibrated spring clamps maintain a constant force of 65 N. In the standard condition, the bonded sample is fully polymerised at 150°C for 3 hours [1]. The thickness of the adhesive was estimated by measuring with a caliper across the entire assembly and then subtracting the thickness of the titanium alloy and that of the composite, as they are relatively thick. In this study, we will work on two samples that are quite similar except for the epoxy thickness. The thickness of the epoxy film in sample 1 is significantly greater than that of sample 2. The parameters of the samples are given in Table 1, as stated by the manufacturer.

Material	ρ (kg/m^3)	C_L (m/s)	C_{11} (GPa)	Sample 1 d (mm)	Sample 2 d (mm)
TA6V	4430	6060	163	0.994	0.981
Epoxy	1300	2300	6.88	0.135	0.086
Composite	1590	2900	13.4	13.5	13.5

ρ (kg/m^3): density; C_L (m/s): velocity; C_{11} (GPa): stiffness coefficient; d (mm): thickness.

Table 1. Studied trilayer assemblies and associated properties. The measurement of the epoxy thickness is approximate (point measurement using calipers).

3. Experimental study

3.1. Experimental protocol

The experimental study is based on the use of the scanning acoustic microscope (SAM) PVA TEPLA 301 (pulse-echo method), whose operating principle relies on the reflection of acoustic waves at interfaces, due to the change in acoustic impedance. It is a sequential representation system in which a piezoelectric transducer emits a beam of ultrasonic waves that propagates through a coupling liquid, which in this case is water. The beam is scattered by the sample, and the signal is received by the same transducer [4] [5] [6]. As shown in Figure 1, the SAM consists of three main functional units, namely: the acoustic sensor, the electronic unit, the mechanical scanning system, and the display device [7].

Journal of Physics: Conference Series **2904** (2024) 012002 doi:10.1088/1742-6596/2904/1/012002

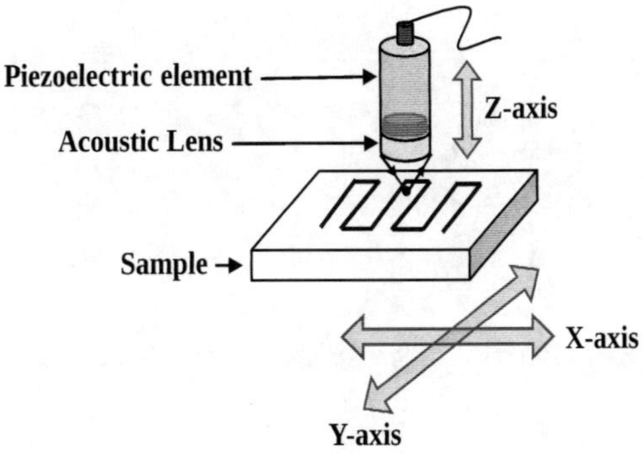

Figure 1. Functional diagram of the scanning acoustic microscope.

It also offers many different imaging techniques. Figure 2 illustrates the various types of scans available; each providing a method of interpretation to visualise and assess the examined material area in a specific way. The three most commonly used scan types with SAM are A-scan, C-scan, and X-scan [8].

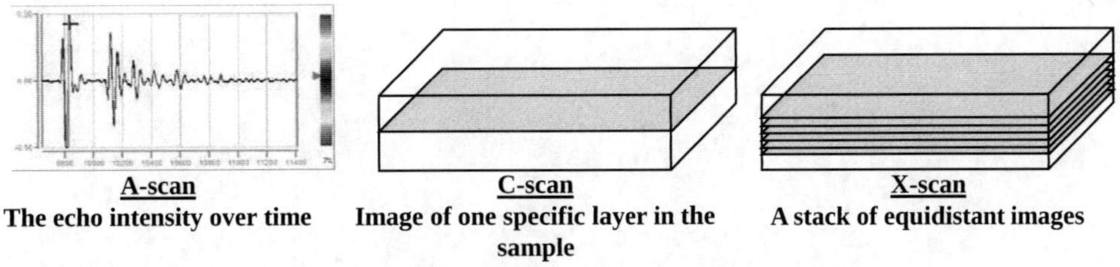

Figure 2. Acoustic imaging modes [8].

3.2. Pulse-echo characterisation using an A-scan signal

Before proceeding with thickness measurements and signal processing, it is essential to check that the platform on which the sample is placed is perfectly level in the X and Y directions. If the platform is not properly adjusted, it can affect the time of flight of ultrasonic waves and lead to errors in signal processing. Additionally, it is important to carefully select the transducer to reduce diffraction effects and to choose the appropriate excitation frequency [9] [10]. Figure 3 and Figure 4 shows the A-scan obtained using a 50 MHz centre frequency planar transducer on the two samples.

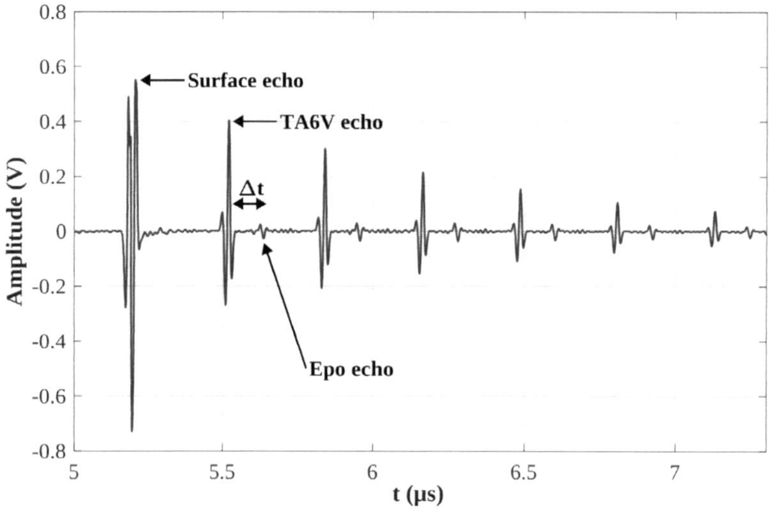

Figure 3. A-scan of sample 1 for 50 MHz centre frequency excitation.

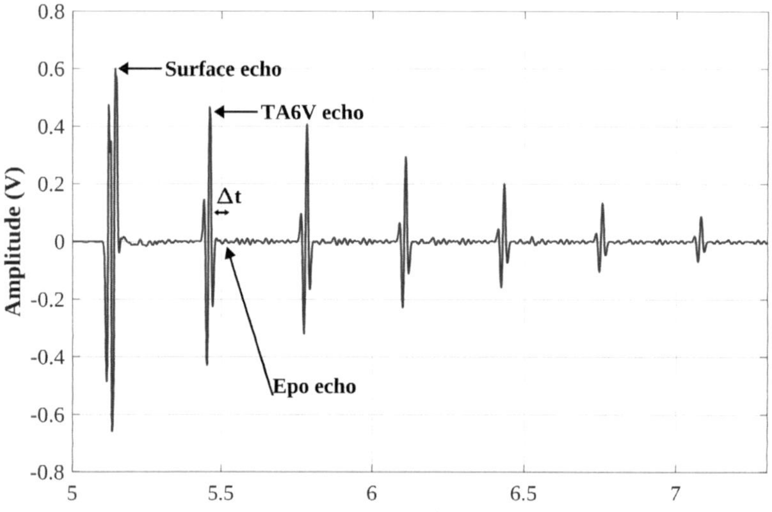

Figure 4. A-scan of sample 2 for 50 MHz centre frequency excitation.

Upon examining the obtained signals, the following can be observed:

- ⇨ the echo at the surface of the Water-TA6V (first interface),
- ⇨ multiple reflections at the TA6V-Epoxy Resin (Epo) interface (second interface),
- ⇨ multiple reflections at the epoxy resin (Epo)-Composite interface (third interface),

The probe-sample distance is approximately 3.85 mm, indicating that the surface echo starts at 5.13 µs. It can be seen that for sample 2, the Epo background echo amplitude is low. Conversely, for sample 1, the epoxy film thickness is significantly greater than that of sample 2, and the Epo background echo amplitude is significant. This leads to the hypothesis that the adhesion quality of sample 2 may be better than that of sample 1, assuming identical cohesive properties of the adhesive.

Accurate measurement of the epoxy film thickness is necessary to ensure the mechanical properties of the adhesive. In practice, strips of a given material are placed around the periphery of the surfaces to be bonded to obtain an adhesive film with the most uniform thickness possible. However, this thickness may change during polymerisation.

Starting from a constant velocity of the adhesive, assuming that it has the same properties for both samples, as stated by the manufacturer, we can evaluate the adhesive thickness. This is done by evaluating the time of flight between the background echo of TA6V and the background echo of the epoxy resin, corresponding to the wave path within the epoxy resin. To do this, we use signal processing methods such as the Hilbert transform and cross-correlation [9]. We first plot the Hilbert transform envelopes for each relevant echo, and then perform a cross-correlation between the two corresponding Hilbert transform envelopes. This method allows us to generate the correlated time of flight Δt_{cor}. The axial resolution of the system at 50 MHz is 23 µm in the Epoxy. Figure 5 illustrates the cross-correlations used to calculate the thickness of the adhesive.

Sample 1: $e_{epoxy} = c_L \times \Delta t_{cor}/2$ = 132.25 µm
Sample 2: $e_{epoxy} = c_L \times \Delta t_{cor}/2$ = 69 µm

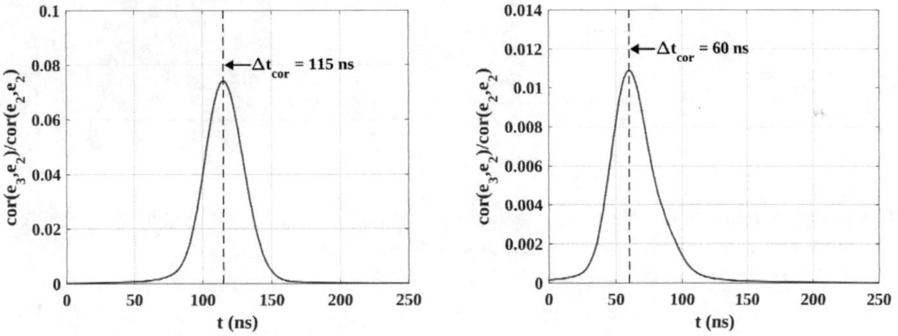

(a) (b)

Figure 5. Cross-correlation of the Hilbert transform envelopes between the background echo of TA6V and the background echo of the epoxy for sample 1 (a) and sample 2 (b).

3.3. Quantification with X-scan imaging

❖ Sample 1

Using a 20 MHz focusing transducer, a X-scan is initiated to obtain 2D images (C-scans) reflecting the subsurface condition. Initially, it is essential to configure the experimental conditions, including setting the correct focal distance (with a focus fixed on the epoxy background echo), calibration, and establishing the protocol for horizontal adjustment. A data acquisition gate is then defined with a duration of approximately 452 ns for sample 1, which is then subdivided into smaller X-scan gates of approximately 20 ns, corresponding to a propagation distance of 23 µm in the epoxy and 29.2 µm in the composite. The axial resolution of the system at 20 MHz is 57 µm in the Epoxy.

As a constitutive element of a X-scan, each C-scan is mapping the amplitude of the echo for a given range of time-of-flight values [11]:

$$s_{C,n}(x,y) = \max_{t_{gate,n} < t \leq t_{gate,n} + \Delta t_{gate}} \{s(x,y,t)\} \quad (1)$$

In the following, each C-scan map denoted "$s_{C,n}(x,y)$" will be referred to as "C-scan *n*", where the number *n* corresponds to each time gate, ranging from $t_{gate,n}$ to $t_{gate,n} + \Delta t_{gate}$ where $\Delta t_{gate} = 20$ ns.

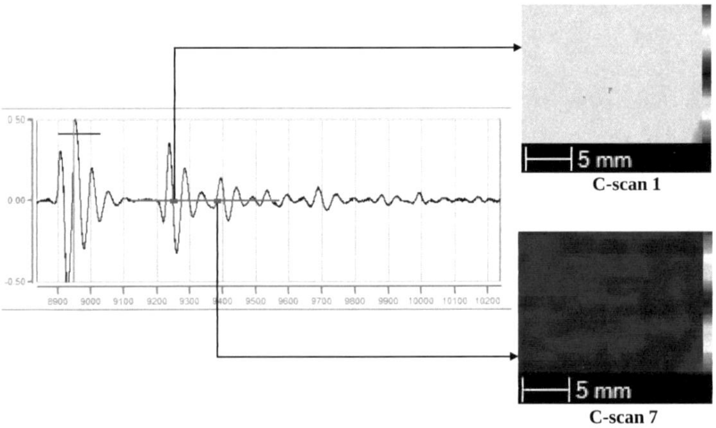

Figure 6. Focused A-scan at 20 MHz with a data gate in red and an X-scan gate in blue, and examples of the corresponding images.

Journal of Physics: Conference Series **2904** (2024) 012002 doi:10.1088/1742-6596/2904/1/012002

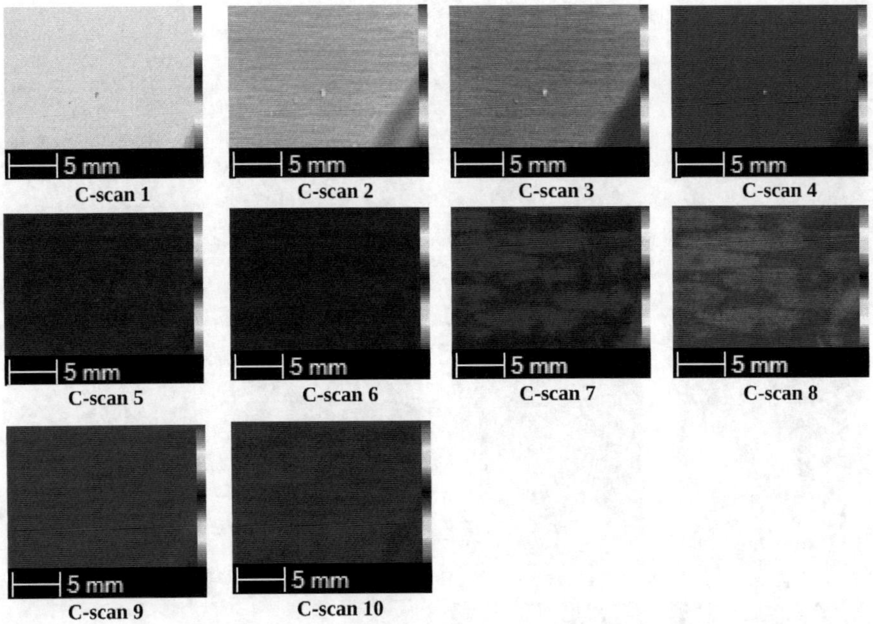

Figure 7. Images obtained from the X-scan of sample 1, starting from the TA6V-Epoxy interface, each cover a surface area of 2×1.5 cm^2, with a pixel resolution of 200 µm/Pixel.

Analysing the images resulting from the X-scan starting from the TA6V-Epoxy interface, uniform colours are observed across most of the image, with some contrast at the edges, possibly indicating a sample misalignment problem. This colour homogeneity suggests that we are still in the same medium, the epoxy. Proceeding through the images, we reach (C-scan 7), where we observe the appearance of carbon fibre bundles of approximately 2.2 mm wide in light blue, mixed with the organic matrix epoxy in dark blue (Figure 8). At this point, we can affirm that we have passed through the epoxy layer and entered the composite zone. At this stage, we can estimate the thickness based on the time shift between C-scan 1 and C-scan 7, which corresponds to 6 windows of 20 ns each. As a result, the thickness of the epoxy layer is estimated to be, e_{epoxy} = 6×23 = 138 µm.

Figure 8. Enlarged image of C-scan 7 of the composite material from sample 1 showing the carbon fibre bundles of 2.2 mm wide in lighter blue mixed with the epoxy in darker blue.

❖ Sample 2

The experimental adjustment protocol is the same as that for sample 1.

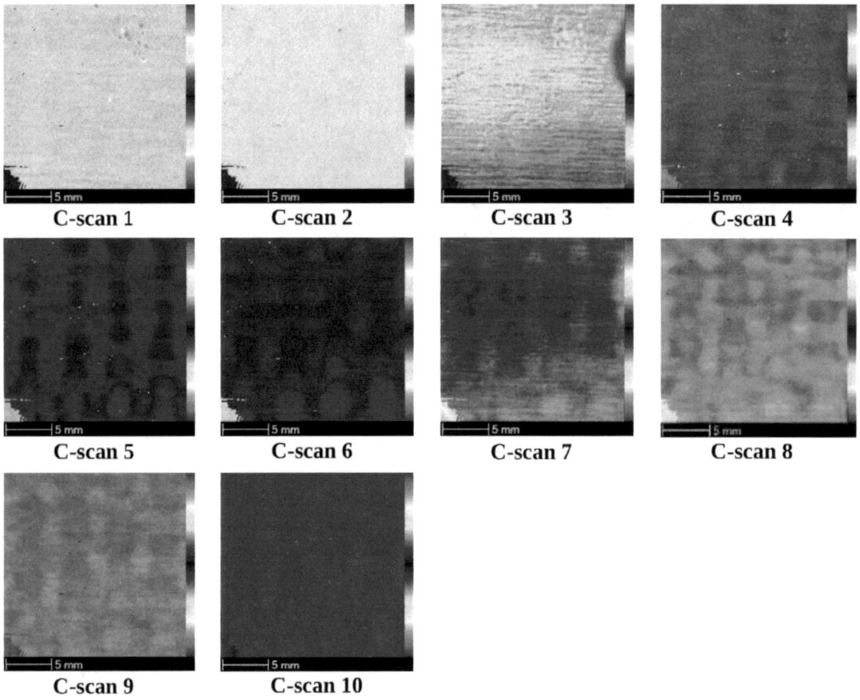

Figure 9. Images obtained from the X-scan of sample 2, starting from the TA6V-Epoxy interface, each cover a surface area of 2×2 cm^2, with a pixel resolution of 100 µm/Pixel.

As in the previous case of study, when analysing the images resulting from the X-scan starting from the TA6V-Epoxy interface, we observe colour uniformity over most of the image, indicating that we are inside the epoxy. Following the evolution of the images (Figure 9), we reach C-scan 4, where we observe the appearance of carbon fibre bundles of approximately 3 mm in light blue, mixed with the epoxy in darker blue. At this stage, we can affirm that we have left the epoxy layer to enter the composite zone; these bundles are very visible and present in many windows. At this point, we can estimate the thickness of the epoxy film based on the time shift between C-scan 1 and C-scan 4, which is 3 windows of 20 ns each. Thus, the thickness of the adhesive epoxy layer is estimated to be, $e_{epoxy} = 3 \times 23 = 69$ µm.

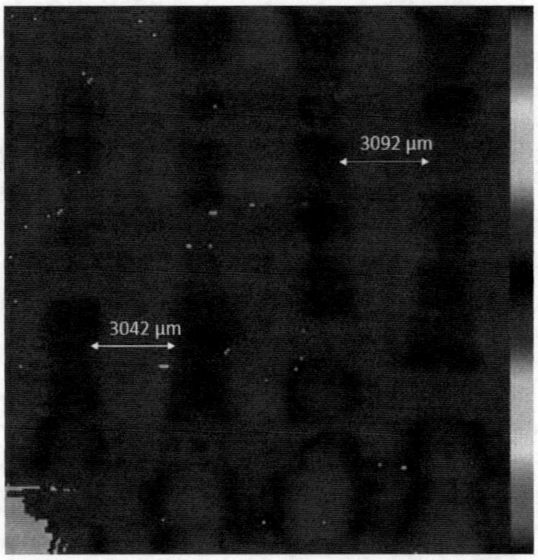

Figure 10. Enlarged image of C-scan 5 of the composite material from sample 2 showing the carbon fibre bundles of 3 mm wide in lighter blue mixed with the epoxy in darker blue.

The Table 2 below summarises the thickness of the epoxy film calculated by A-scan and X-scan imaging.

Measurement method	Measured Thickness for Sample 1 (µm)	Measured Thickness for sample 2 (µm)
Estimated by Safran: Caliper cross	135	86
X-scan Imaging	138	69
A-scan	132.25	69

Table 2. Measured thicknesses of the epoxy film samples by A-scan and X-scan.

Using A-scan and X-scan imaging methods, we were able to accurately determine the thickness of the adhesive acoustically. It is at 132.25 µm according to the A-scan, and at 138 µm according to the X-scan imaging. For sample 2, the measured thickness of the epoxy film using both methods is approximately 69 µm.

4. Results and discussions

Initially, several samples with different adhesive film thicknesses were tested using flat or focused transducers. During these pulse-echo measurements, it was observed that only certain samples, with adhesive film thicknesses less than 200 µm, exhibited round-trip echoes. This observation is consistent with the fact that the adhesive is highly attenuating, leading to a further reduction in the energy reaching the epoxy-composite interface as the adhesive thickness increases. In addition, it was necessary to select the samples in a manner that avoided overlap between the echoes from TA6V and those from the adhesive film, thus facilitating subsequent signal processing. Ultimately, two samples were selected for this study. Furthermore, accurate measurement of the epoxy film thickness is essential to ensure the mechanical properties of the adhesive. In practice, strips of material placed on the edges of the surfaces between the two materials during the bonding process allow for the thickness to be controlled. However, this thickness can change during polymerisation. Therefore, the thickness of the epoxy film after the bonding process is only an estimate that must be validated by acoustic methods. The measurement of the epoxy film thickness for sample 1 is at 132.25 µm according to the A-scan, and 138 µm according to the X-scan imaging. For sample 2, the measured thickness of the epoxy film using both methods is approximately 69 µm.

In terms of adhesion quality, and by comparing the results of the two samples, it can be seen that for sample 2, with an epoxy film thickness of 69 µm, the background echo amplitude is low. Conversely, for sample 1, the epoxy film thickness is significantly greater than that of sample 2, and the Epo background echo amplitude is significant. This difference in amplitude cannot be attributed to the attenuation due to the viscosity of the adhesive, but rather to the interface conditions, given the internal attenuation of the adhesive. We can conclude that since the reflected echo exhibits a higher amplitude in sample 1 compared to sample 2, the level of adhesion could be better in sample 2 than in sample 1. In addition, the composite is clearly visible in several windows of the X-scan imaging for sample 2 (6 C-scan windows, from C-scan 4 to C-scan 9), but it is less visible and present in only a few windows of the X-scan imaging for sample 1 (2 C-scan windows, from C-scan 7 to C-scan 8). These observations highlight the energy transmission within the composite and enable the differentiation of carbon fibres with varying clarity through different windows. The number of images, which is a function of the distance travelled within the composite, confirms that the level of adhesion is indeed better in sample 2 than in sample 1.

5. Conclusion

Two trilayer samples were studied to evaluate the properties of the epoxy film. The challenge lies in the very contrasting impedance discontinuities in this type of structure: strong between the titanium and the adhesive, and weak between the adhesive and the composite. These samples were then characterised using the PVA TEPLA SAM 301 scanning acoustic microscope. One of the objectives was to evaluate the thickness of the epoxy film. Quantification was also carried out using X-scan imaging with a 20 MHz focusing transducer to obtain various C-scans at different depths, particularly from the epoxy to the composite layers, also allowing measurement of thickness through imaging. Different adhesion qualities of the samples were identified by analysing the background echo of the epoxy film relative to its thickness, and by examining the amount of energy transmitted to the composite layer through the X-scan imaging.

Acknowledgment

The authors would like to thank the SAFRAN group (Mathieu Ducousso and Nicolas Cuvillier) for providing the samples.

References

[1] R. Hodé, S. Raetz, J. Blondeau, N. Chigarev, N. Cuvillier, V. Tournat and M. Ducousso, "Nondestructive evaluation of structural adhesive bonding using the attenuation of zero-group-velocity Lamb modes," *Applied Physics Letters,* vol. 116, p. 104101, 2020. https://doi.org/10.1063/1.5143215

[2] M. Ducousso, S. Bardy, Y. Rouchausse, T. Bergara, F. Jenson, L. Berthe, L. Videau and N. Cuvillier, "Quantitative evaluation of the mechanical strength of titanium/composite bonding using laser-generated shock waves," Applied Physics Letters, vol. 112, p. 111904, 2018. https://doi.org/10.1063/1.5020352

[3] L. Attar, M. Ech Cherif El Kettani, D. Leduc, M. V. Predoi and J. Galy, "Detection of kissing bond type defects and evaluation of the bonding quality in metal/adhesive/composite structures by a wavenumber-frequency insensitive SH mode," *NDT & E International,* vol. 137, p. 102841, 2023. https://doi.org/10.1016/j.ndteint.2023.102841

[4] R. A. Lemons and C. F. Quate, "Acoustic microscope - Scanning version," Applied Physics Letters, vol. 24, p. 163–165, 1974. https://doi.org/10.1063/1.1655136

[5] R. A. Lemons and C. F. Quate, "Integrated circuits as viewed with an acoustic microscope," *Applied Physics Letters,* vol. 25, p. 251–253, 1974. https://doi.org/10.1063/1.1655459

[6] V. Jipson and C. F. Quate, "Acoustic microscopy at optical wavelengths," Applied Physics Letters, vol. 32, p. 789–791, 1978. https://doi.org/10.1063/1.89931

[7] F. Hamdi, S. Bouhedja and H. Amrani, "Theoretical study of different attenuation measurement by acoustic microscopy," *Journal of Applied Physics,* vol. 114,p. 133501 ,2013. https://doi.org/10.1063/1.4823850

[8] F. Bertocci, A. Grandoni and T. Djuric-Rissner, "Scanning Acoustic Microscopy (SAM): A Robust Method for Defect Detection during the Manufacturing Process of Ultrasound Probes for Medical Imaging," Sensors, vol. 19, p. 4868, 2019. https://doi.org/10.3390/s19224868

[9] N. Ghodhbani, P. Marechal and H. Duflo, "Ultrasonic broadband characterization of a viscous liquid: Methods and perturbation factors," Ultrasonics, vol. 56, pp. 308-317, 2015. https://doi.org/10.1016/j.ultras.2014.08.013

[10] N. Samet, P. Maréchal and H. Duflo, "Ultrasonic characterization of a fluid layer using a broadband transducer," Ultrasonics, vol. 52, pp. 427-434, 2012. https://doi.org/10.1016/j.ultras.2011.10.004

[11] Naim Samet, Pierre Marechal, Hugues Duflo. Ultrasonic imaging of bubble motion in a fiber preform. Acoustics 2012, pp. 2601- 2606, 2012, Nantes, France. hal-00810855

Journal of Physics: Conference Series **2904** (2024) 012003 doi:10.1088/1742-6596/2904/1/012003

Ultrasonic characterization of post-mortem interval (PMI) of human bones

Arthur Angermuller [a], Hafsa Diboune [a], Andres Arciniegas [a*], Loïc Martinez [a], Philippe Lasaygues [b], Cécile Baron [c], Nicolas Wilkie-Chancellier [a]

a Laboratoire SATIE, UMR CNRS 8029, CY Cergy Paris Université, 5 mail Gay Lussac, 95031 Neuville sur Oise, France
b Aix-Marseille Univ, CNRS, Centrale Méditerranée, LMA, Marseille, France
c Aix-Marseille Univ, CNRS, Centrale Méditerranée, IRPHE, Marseille, France

* Corresponding author email: andres.arciniegas-mosquera@cyu.fr

Abstract. In order to improve the evaluation of the post-mortem interval (PMI) of human bone, experts in forensic anthropology from the french national gendarmerie are interested in new characterization methods and in particular non-destructive testing (NDT). If Nile Blue colorimetric methods currently allow dating up to a hundred years, these techniques alter the bone which is evidence in a criminal investigation. In order to avoid this degradation, the work presented hereby proposes a method for characterizing PMI using ultrasonic methods that have proven successful for the NDT of complex materials, including bones in medical applications. The objective is to identify relevant ultrasonic parameters representative of PMI. The propagation velocities of compression and shear waves are measured through parallelepiped samples of cortical bone taken from human femurs while respecting the anatomical orientation of the bone. These measurements carried out in the 3 directions of space make it possible to calculate the diagonal coefficients of the stiffness matrix Cij. Measurements of the propagation velocity of compression waves are carried out in transmission, using Pinducer™ transmitter/receiver immersed in water. The measurement of shear wave velocities is made via a second non-immersed device using contact transducers. The results presented come from ultrasonic signals measured on bones, coming from individuals with similar pre-mortem parameters, whose PMI varies between 1 and 2500 years. The six diagonal coefficients of the stiffness matrix are represented and discussed in terms of the PMI. A first classification of bones is possible according to their PMI. In order to assist the analysis and interpretation of the results, a single parameter is proposed, the Trace of the stiffness matrix (Tr(Cij)), confirming the possible contribution of the ultrasonic techniques to be used for the dating of human bones.

1. Introduction

Bone is a particularly complex material to study because of its internal structure, its geometry and the differences between individuals. Many methods make it possible to study the degradation of bone over time after the death of an individual (i.e. phenomenon called diagenesis). These methods can be [1]:

biochemical estimation of the level of organic elements still present in bone; radioisotope measurement of levels of unstable chemical components still present in the bone; and histological study of the internal structural degradation of bone.

In the case of forensic anthropology, when a crime scene investigation is opened, a quantity of individualizing information is collected on the victim to define their biological profile and ultimately obtain information on the criminals. When only the bones remain of the victim (in particular cortical bone), one of the fundamental pieces of information is the estimation of the post-mortem interval (PMI): the time between the death of the individual and the discovery of the body. This interval determines the legal outcome or not of most of investigations: when PMI exceeds the limitation period (6-20 years), a court will not accept evidence according to the french Law. To determine PMI, thanatology and entomology methods can be used for intervals less than 1 year, but beyond, colorimetric methods and radioisotope methods should be favored. However, the methods accuracy and precision have limits due to the multiple parameters influencing the determination of PMI, such as pre-mortem parameters of the individual (age, sex, physical state, illness) and environmental parameters of place of discovery (humidity, temperature, burial, geological zone).

The French national gendarmerie forensic department (IRCGN) was interested in new characterization methods and in particular non-destructive testing (NDT). Since Nile Blue colorimetry methods currently allow dating up to a hundred years, these techniques alter the bone which is evidence in a criminal investigation [2]. In order to avoid this degradation, the work presented hereby proposes a method for characterizing PMI using ultrasonic techniques that have proven successful for the NDT of complex materials [3], including bones in medical applications [4].

The motivation of this study is that mechanical behavior depends on composition and structure of the material. Bone mechanical behavior requires the use of hypotheses to reduce the number of parameters in order to approach the bone using simple models. Many works have proposed to circumvent the problems linked to the geometry of the bone by studying small samples [5-7]: these studies attempt to approach the mechanical behavior of the bone by considering the medium as infinite. The mechanical behavior of the bone is thus associated with an anisotropy model whose parameters are the stiffness coefficients.

In this paper, we propose to study the stiffness coefficients of bone and their behavior according to PMI. These coefficients will make it possible to approach the behavior of the bone by an anisotropy model and to judge the relevance of this model in the post-mortem time. Independently of the model, the evolution of the measured coefficients will allow us to know if the degradation of the bone matrix linked to diagenesis allows a distinction between different groups of individuals with similar PMI.

This study is hereby presented as follows. Firstly the theoretical background regarding the ultrasonic wave propagation in cortical bone will be presented. Only compression and shear waves were assumed to be transmitted in the sample (direct transmission with no reflection or diffraction). The wave velocities were assumed to be independent of the frequency over the frequency range studied and the waves attenuations were neglected. Secondly, the experimental technique used to study samples with different PMI is presented. Finally, results will be presented and the correspondences and discrepancies are discussed. In order to assist the analysis and interpretation of the results, the Trace of the stiffness matrix ($\mathrm{Tr}(C_{ij})$) is used as unique parameter confirming the possible contribution of the ultrasonic techniques for the dating of human bones.

2. Theoretical models of ultrasonic wave propagation in cortical bone

In this section, the theory behind the proposed non-destructive evaluation method is presented, starting with the different models of bone anisotropy and the study of the ultrasonic wave propagation in homogeneous solids.

In this study, we focused on on the diaphyseal/cortical zone of the human long bone considered as an elastic medium (inhomogeneous and anisotropic). Pre-mortem human bones are mainly Haversian, and which are usually considered to be transversely isotropic [5-7]. However, we will investigate post-mortem bone in order to check if this behavior is conserved during the aging after death and its homogeneity.

The notion of anisotropy describes a material whose mechanical behavior varies depending on the axis of study and the nature of the stress applied. These theoretical notions being widely presented in the literature [8-10], only a brief reminder is proposed allowing us to introduce the elements necessary for understanding anisotropy.

The objective of mechanical monitoring of a material is to predict its evolution when it is subjected to stress. Hooke's law relates the stress T_{ij} applied to a solid material to the strain S_{kl} through the elastic stiffness tensor C_{ijkl} by the following formula:

$$T_{ij} = C_{ijkl} S_{kl} \quad (1)$$

The index *ijkl* are integer numbers from 1 to 3 according to the axes (*x, y, z*) of geometric system in Cartesian coordinates. Thus, the maximum possible number of stiffness constants in Hooke's law is $3^4 = 81$.

The displacement of matter following ultrasonic stimulation remains within the elasticity limit: the material can return to its initial state after the loading cycle. This remark allows us to define the system as being in static equilibrium, ensuring the condition of motion equilibrium. This condition forces the symmetry of the tensors *T* and *S*, allowing us to use Einstein's notation and thus simplify the writing of Hooke's law:

$$\begin{matrix} 11 & 12 & 13 \\ 21 & 22 & 23 \\ 31 & 32 & 33 \end{matrix} = \begin{matrix} 1 & 6 & 5 \\ 6 & 2 & 4 \\ 5 & 4 & 3 \end{matrix} \quad (2)$$

This transformation implies that Hooke's law can then be written:

$$T_i = C_{ij} S_j \quad (3)$$

The index *i* and *j* being between 1 and 6, the tensor C_{ij} is represented in the form of a matrix of 6 columns and 6 rows and is composed of 36 coefficients. This number of coefficients can be reduced to 21 thanks to the symmetric character of the tensor corresponding to the triclinic system. On the other hand, the organization of anatomical elements explains anisotropic behavior of cortical bone. Thus, the degree of anisotropy of the cortical bone makes it possible to approach it by a material with orthotropic symmetry [11], where *z* is denoted as the axis of symmetry.

Taking into account orthotropic symmetry, the tensor can then be reduced to 9 coefficients:

$$Orthotropic\, C_{ij} = \begin{bmatrix} C_{11} & C_{12} & C_{13} & 0 & 0 & 0 \\ C_{12} & C_{22} & C_{23} & 0 & 0 & 0 \\ C_{13} & C_{23} & C_{33} & 0 & 0 & 0 \\ 0 & 0 & 0 & C_{44} & 0 & 0 \\ 0 & 0 & 0 & 0 & C_{55} & 0 \\ 0 & 0 & 0 & 0 & 0 & C_{66} \end{bmatrix} \quad (4)$$

Modeling as a transversely isotropic elastic solid material [5-7,12,14] and isotropic were also considered [13] allowing reduction to 6 or 3 coefficients.

$$\text{Transversely isotropic}\, C_{ij} = \begin{bmatrix} C_{11} & C_{12} & C_{13} & 0 & 0 & 0 \\ C_{12} & C_{11} & C_{13} & 0 & 0 & 0 \\ C_{13} & C_{13} & C_{33} & 0 & 0 & 0 \\ 0 & 0 & 0 & C_{44} & 0 & 0 \\ 0 & 0 & 0 & 0 & C_{44} & 0 \\ 0 & 0 & 0 & 0 & 0 & C_{66} \end{bmatrix} \tag{5}$$

The study of these matrices can be done by monitoring certain parameters. In particular, the choice of model can be defined using anisotropy ratios [15]. In the case of the transversely isotropic model, the expected C_{11}/C_{22} and C_{44}/C_{55} ratios are close to 1.

Considering ultrasonic wave perturbation of the material, combining the motion conservation equation, Hooke's law, and the kinematic relations between the strains produced by the particle displacement due to wave propagation, the wave equation can be deduced [10]. The progressive plane wave solution leads to the Christoffel's equation for which the phase velocities are solutions to an eigenvalue problem. The diagonal coefficients of the stiffness matrix can then be associated with the propagation phase velocities with Einstein's notation:

$$C_{ij} = \rho V_i^2 \tag{6}$$

with ρ the mass density. With the knowledge of this last quantity and measuring the velocities of pure compressional and shear waves propagating along the three principal axes, the diagonal elements of the stiffness matrix, can be therefore computed, as well as the Trace of the stiffness matrix and the anisotropy ratios.

3. Experimental technique

In this section, the experimental technique is presented, starting with the experimental protocol, following samples conditioning, the description of the ultrasonic measurement and finishing with the hypotheses to be checked. The measurement technique is based on classical approach for the analysis of ultrasonic echoes in locally parallel samples.

Experimental protocol

The experiments were performed on 11 individual femur samples belonging to 4 different post-mortem interval (PMI) groups. Each individual sample was cut into cubes (3 cubes per individual) in order to carry out the ultrasonic measurements. The dimensions of samples are sized with digital caliper and mass density is controlled using a weighing machine and a density kit.

Ultrasonic waves measurements are carried out in transmission mode following the three axes of the cubic samples. Compression waves are measured using transducers immersed in water, while shear waves are measured via a second non-immersed device using contact transducers. All received ultrasonic signals were recorded using a digital oscilloscope and the propagation velocities were determined using propagation path length and time-of-flight detection. Finally the diagonal elements of the stiffness matrix are computed from the mass density and the velocities.

Journal of Physics: Conference Series **2904** (2024) 012003 doi:10.1088/1742-6596/2904/1/012003

Samples conditioning

The human bones delivered by the IRCGN have already been cleaned and washed before their arrival. This work focused on samples of cortical bone from adult human femurs. Pre-mortem parameters (age, sex, physical state, illness) allowing determination of the identity of individual and exact circumstances of their discovery are unknown. However, the samples provided come from similar discovery environments according to the IRCGN.

The set of human bones studied is composed of 11 individuals of different PMI: three from 1 to 3 years (Group 1), three from 10 to 13 years (Group 2), three from around 50 years (Group 3) and two of archaeological ages (Group 4). The group 1 is considered as control group (similar to medical studies), group 2 has PMI close to the limitation period, group 3 has PMI beyond limitation period and group 4 represents extreme situation. The sample summary is shown in Table 1.

Table 1: Summary of PMI groups and porosity measurements.

Group 1 (G1)	1 year (ϕ = 5.8 %)	3 years (ϕ = 10.7 %)	3 years (ϕ = 5.2 %)
Group 2 (G2)	10 years (ϕ = 5.3 %)	11 years (ϕ = 11.6 %)	13 years (ϕ = 7.9 %)
Group 3 (G3)	~50 years (ϕ = 6.4 %)	~50 years (ϕ = 4.1 %)	~50 years (ϕ = 5.6 %)
Group 4 (G4)	500 years (ϕ = 6.1 %)	2500 years (ϕ = 5.0 %)	

X-ray micro-tomography scans on the bones entrusted by the IRCGN were carried out in the Laboratory of Mechanics and Acoustics of Marseille before their cutting in cube. Each individual bone was cut into three parallelepipeds (hereinafter named cube) of dimension (3-6 x 5 x 5 mm): the X-ray images made it possible to verify the porosity of the samples (Table 1). Even if the dimensions between individuals vary, the cutouts sought to maximize the dimensions of the sides, particularly in the radial axis, in the cortical thickness: these dimensions are between 3 and 6 mm. Some samples, which had thin cortical thickness and/or had high porosity, made difficult data analysis (in red in Table 1): for these samples in particular, other cubes were cut at different locations on the bone.

The cuts were made using a diamond saw (IsoMet™ Low Speed Saw, Buehler, Lake Bluff, IL, USA) with a slow rotation, following directions shown in Figure 1A. This saw, particularly suited to cutting fragile samples, makes it possible to hold the sample and apply weight to it when cutting to control the pressure on the saw. The rotating blade is bathed in an oil bath to limit friction with the sample when cutting. Finally, the arm holding the sample can be shifted above of the blade without having to touch the sample, thus ensuring the parallelism of the opposite faces of the cube.

The three cubes from each individual are cut one after the other along the axis of the bone fibers: the samples allow to perform three measurements (on the three cubes) on a length of ~1.5 cm (3 x 0.5 cm) along the axis of the fibers (Figure 1B). This cutting protocol allows to question the local homogeneity of the bone along the fibers.

Before carrying out compression waves measurements, the cubic samples were degassed (also allowing measurement of mass density using the Archimedes' principle), while for the shear wave measurements the samples were dried to facilitate the wave propagation. The mass density was therefore obtained using a weighing machine (Ohaus PX323M) and a density kit (Voyager 610, Ohaus Corporation, FlorhamPark, NJ, USA).

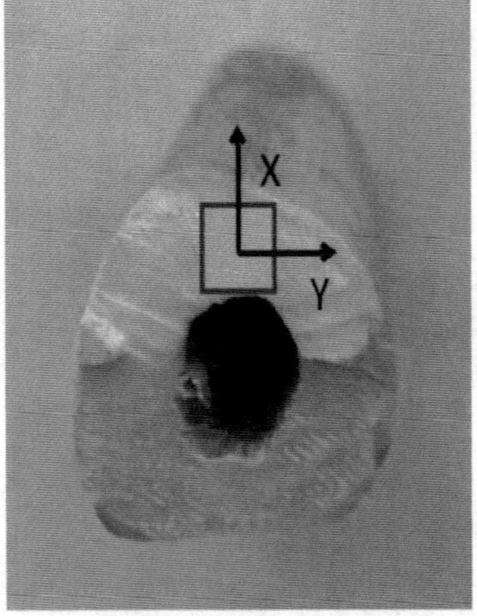

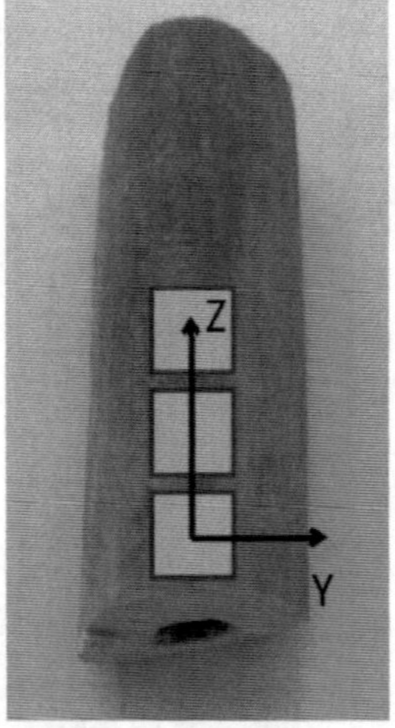

Figure 1: Bone sample conditioning. A) XY plane view. B) YZ plane view.

Ultrasonic measurements

Figure 2 shows the principle of measurement of ultrasonic velocities V_{ii} of locally parallel sample. The experiments were carried out in the transmission mode between two coaxial transmitter/receiver transducers (5 MHz Pinducer^{TM}, CTS Valpey Corporation, Hopkinton, MA, USA) as described in [15]. Each transducer was driven using a pulse-receiver generator (Sofranel 5077 PR, Olympus, USA) including the voltage amplifier. The radio-frequency signals (RF-signals) were conveyed from the oscilloscope (Agilent DSO1514A, Santa Clara, CA 95051, USA) to a personal computer using a USB interface file transfer, and stored. All processing algorithms were implemented using Matlab (The MathWorks, Inc., Natick, Massachusetts, USA).

In order to operate with well-defined coupling conditions an immersion technique was used [16-18]. Measurements were performed, one without the sample (reference measurement) and the other with the sample. This procedure was repeated five times per cube axis. The energy delivered by the transmitting transducer and the distance to the receiving transducer remain invariant for both measurements.

In order to calculate the last three diagonal coefficients of the stiffness matrix, the shear wave velocities were measured. These waves propagate poorly in the water, a contact measurement is necessary. To measure the propagation velocities shear waves in the three planes, a test bench was created: as for the measurements in immersion, this bench has already been used to characterize children's bones [15]. As in immersion technique, the experiments were carried out in the transmission mode between two coaxial transmitter/receiver transducers (2.25 MHz Panametrics V154, Olympus, Waltham, Massachusetts, USA, Figure 3) [19]. Measurements were performed, one with a reference sample (i.e. a homogeneous and isotropic parallelepiped aluminum sample, 3 mm thick) and the other with the bone sample. The coupling medium used between the transducers and the samples was a viscous Shear Waves Coupling suitable for transverse stimulation unlike aqueous coupling. The shear

Journal of Physics: Conference Series **2904** (2024) 012003 doi:10.1088/1742-6596/2904/1/012003

measurements were made according to a plane: an axis of propagation *i* and an axis of polarization *j* for the velocity V_{ij}. In total, six velocity measurements can be made according to the three axes and reversing the direction of propagation and polarization.

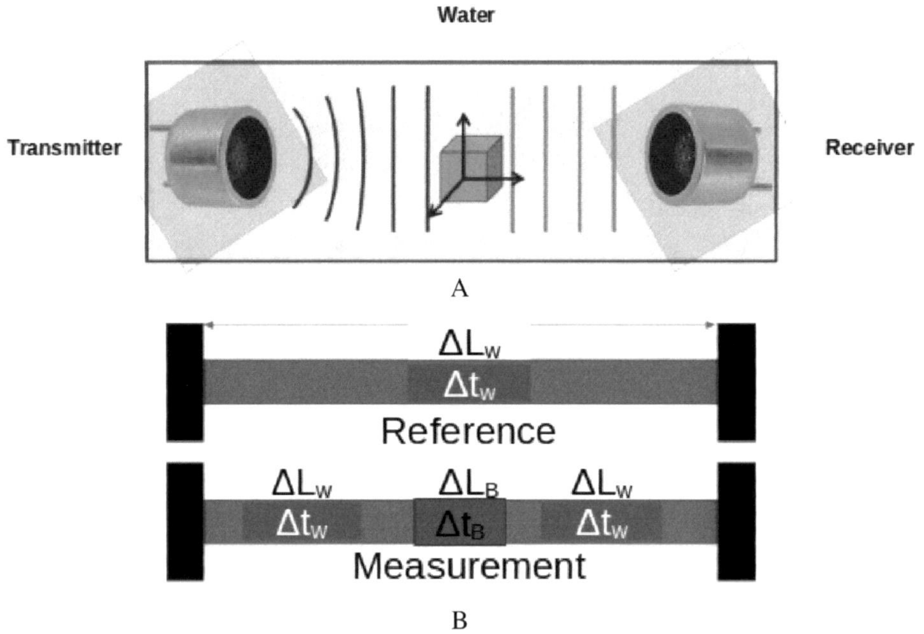

Figure 2: Immersion method. A) Principle. B) Definition of distances and propagation times.

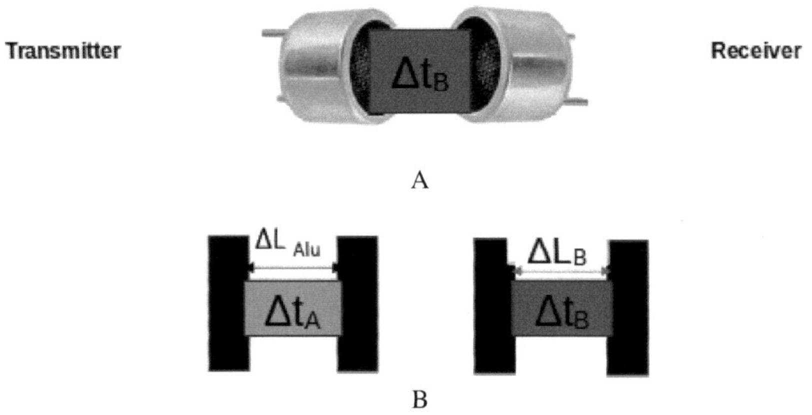

Figure 3: Contact method. A) Principle. B) Differential measurement using reference aluminum and bone samples.

For the frequency range studied, the average wavelength for longitudinal and shear waves in human cortical bone is approximately of ~0.8 mm, which let us to study bone structure at mesoscopic scale. This wavelength is sufficient to avoid echo overlapping according to our samples sizes, and therefore to be able to compute wave velocities using temporal signal processing. Thus, time-of-flight difference between measurement with bone sample and reference sample (denoted Δt), is determined by cross-correlation of the recorded RF-signal. Finally, compression and shear waves velocities are determined using equations 7 and 8 respectively and diagonal coefficients of stiffness matrix can be deduced using equation 6.

$$V_{ii}=\frac{V_{water}}{1-\frac{\Delta t_{ii}}{\Delta L_B}V_{water}} \tag{7}$$

$$V_{ij}=\frac{\Delta L_B}{\frac{\Delta L_{alu}}{V_{alu}}-\Delta t_{ij}} \tag{8}$$

Assumptions to be checked

Even if the aim of ultrasound measurements is to identify changes in the bone over post-mortem time, many parameters can lead to differences between individuals studied, whether related to PMI or not, known or not. In order to limit the number of parameters monitored, several hypotheses are made about the mechanical behavior of the bone.

The first hypothesis states that local homogeneity of the bone allows us to consider all measurements on bone as equivalent regardless of the cubic sample chosen. The interest of this hypothesis is to question the human factor on the measurements. If the measurements strongly depend on the location of the sample in the cortical bone, then any conclusion between individuals must take into account the anatomical location. Theoretically, if discrepancies of the measured stiffness constants can be considered as low, then the hypothesis is verified.

The second hypothesis states that a transversely isotropic model makes it possible to reduce the number of stiffness coefficients compared to the orthotropic model. It allows firstly to associate C_{11} with C_{22} and C_{44} to C_{55}, and secondly to estimate the coefficient extra diagonal C_{12}. Theoretically, if $C_{11} = C_{22}$ and $C_{44} = C_{55}$ then the hypothesis is checked.

The last hypothesis states that the post-mortem mechanical evolution of the bone would be stronger than the pre-mortem parameters between individuals. This hypothesis suggests the possibility of grouping individuals with PMI of the same order of magnitude, and to find singular differences between these groups. The classification of bones should be possible according to their mechanical behavior. Theoretically, the use of a mechanical parameter that highlights groups of similar PMI should let to check this last hypothesis.

The hypotheses are accepted or rejected using Wilcoxon – Mann – Whitney (WMW) statistical test. This test allows to compare membership of a sample according to two groups of data (low number of samples, <30). The significance level is $p<0.01$ (very strong presumption against the null hypothesis) and $p<0.05$ (strong presumption against the null hypothesis).

4. Results and discussion

In this section, the experimental results are presented and discussed: from the calculation of the diagonal coefficients of the stiffness matrix, followed by the calculation of the Trace of the stiffness matrix and the anisotropy ratios.

Diagonal coefficients

In this part we present the evolution of diagonal coefficients of the stiffness matrix as function of the post-mortem interval (PMI). The diagonal coefficients of the stiffness matrix were computed using the mass densities and compression and shear wave velocities measurements according to the three main directions of bone samples. Firstly, the last three coefficients resulting from the shear measurements were calculated using the averages of the corresponding velocities V_{ij} and V_{ji}. In order to ensure the comparability of individuals with each other, the two individuals presenting high porosity rates (> 10%) were removed from the data: the selected individuals then have an average porosity rate of 5.8% with a maximum standard deviation of 2.1%.

Figure 4 presents the diagonal coefficients measured for all individuals according to their PMI. For each group, two zones were delimited, letting to separate samples with high and low values of stiffness coefficients C_{ii}. For all $C_{ii,}$ the highest values were obtained for the samples with PMI less than 50 years and is highlighted that C_{ii}, may decrease between PMI ranging from 13 to 50 years (between G2 and G3). Ultrasound measurements on individuals of different PMI were therefore compared with each other without taking into account external data presented in the literature.

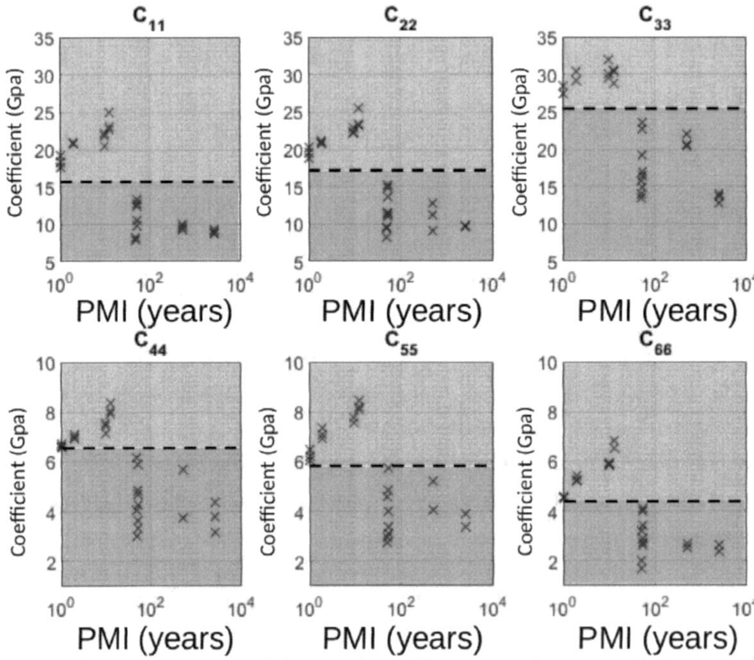

Figure 4: Measured diagonal coefficients as function of PMI.

Journal of Physics: Conference Series **2904** (2024) 012003 doi:10.1088/1742-6596/2904/1/012003

The coefficient values presented show a slight increase between groups G1 and G2 (from 1 to 3 GPa), then a drastic drop between groups G2 and G3 (close to a factor of 2). Groups G3 and G4 do not seem to differ. These differences between groups are highlighted by the WMW statistical test. The 0.05 threshold makes it possible to distinguish groups G1 // G2 and G2 // G3 as different populations (p-value < 0.04 for all C_{ii}). On the other hand this threshold does not allow the G3 // G4 groups to be dissociated (agreement with the null hypothesis, p-value > 0.21 for all C_{ii}).

The precision of measurement of the diagonal coefficients per individual is shown in Table 2. Overall precision per diagonal coefficient ranges between 3% (C_{11}) to 15% (C_{44}). However, the standard deviation of the measurement of C_{44}, C_{55} and C_{66} can be greater than 20% according to the individual. The average precision per individual ranges from 2% to 20%, with the lowest values for the individuals of groups G1 and G2 (with PMI lower than 50 years). Therefore, this leads to validate local homogeneity assumption for individuals of G1 and G2, whereas is less valid for G3 and G4. For these last, the measurement of the shear waves and therefore the associated coefficients may be impacted due to the heterogeneity of those samples.

Table 2: Precision of diagonal coefficients measurement.

	Sample	**C_{11}**	**C_{22}**	**C_{33}**	**C_{44}**	**C_{55}**	**C_{66}**	**Average per individual**	**Average per PMI group**
G1	Ind 1	4%	4%	2%	1%	3%	1%	**3%**	**2.5%**
	Ind 3	1%	0%	2%	1%	3%	2%	**2%**	
G2	Ind 1	4%	2%	4%	3%	1%	1%	**3%**	**3.5%**
	Ind 3	5%	5%	3%	3%	2%	3%	**4%**	
G3	Ind 1	2%	8%	8%	36%	29%	35%	**20%**	**13%**
	Ind 2	3%	5%	11%	28%	10%	10%	**11%**	
	Ind 3	4%	2%	7%	15%	13%	4%	**7%**	
G4	Ind 1	4%	17%	4%	29%	18%	3%	**12%**	**9.5%**
	Ind 2	3%	2%	5%	16%	8%	6%	**7%**	
	Average	**3%**	**5%**	**5%**	**15%**	**10%**	**7%**		

Table 3 presents the average of these coefficients for the four different PMI groups as well as their standard deviations. The standard deviations presented show repeatable measurements for groups G1 and G2 both between individuals of the same PMI, but also between the three samples from the same individual (maximum standard deviation of 8.6% for these two groups). This remark is not valid for groups G3 and G4 whose standard deviations can reach 38%. We recall that the individuals retained for these four groups are assumed comparable: neither their porosity rates nor their densities allow them to be distinguished. The values of measured coefficients for groups G3 and G4 are comparable, not allowing us to differentiate them. The PMI is the only known parameter that can differentiate these individuals. Unlike groups G1 and G2, the measurements of the coefficients depend on one hand, on their different post-mortem alterations, between individuals and on the other hand, between samples from the same individual.

To our knowledge, there are no studies carried out using ultrasound on the question of post-mortem bone degradation. However, the results suggest a late evolution of the coefficients for PMI at least greater than 10 years. This remark suggests that the results obtained on recent individuals are comparable to studies carried out in the medical field: to approach the conditions of living bone, these studies were carried out on fresh bones. The results of groups G1 and G2 of PMI between 1 and 13 years are in agreement with adult bone studies in the medical field [6,15,20].

Table 3: Evolution of diagonal coefficients as function of PMI group.

	C_{11}	C_{22}	C_{33}
G1: 1 – 3 years	19.6 GPa ± 7.2%	20.2 GPa ± 6.3%	28.6 GPa ± 7.0%
G2: 10 – 13 years	22.5 GPa ± 6.4%	23.2 GPa ± 5.4%	30.2 GPa ± 3.6%
G3: ~50 years	10.4 GPa ± 22.7%	11.7 GPa ± 23.5%	17.4 GPa ± 22.1%
G4: Archeo	9.3 GPa ± 5.6%	10.4 GPa ± 8.7%	17.2 GPa ± 31.1%
	C_{44}	C_{55}	C_{66}
G1: 1 – 3 years	6.8 GPa ± 3.9%	6.7 GPa ± 7.6%	4.9 GPa ± 8.6%
G2: 10 – 13 years	7.7 GPa ± 5.6%	8.0 GPa ± 4.1%	6.3 GPa ± 7.5%
G3: ~50 years	4.1 GPa ± 37.3%	4.0 GPa ± 38.4%	2.9 GPa ± 38.8%
G4: Archeo	4.5 GPa ± 17.7%	4.0 GPa ± 17.6%	2.6 GPa ± 4.7%

Trace of stiffness matrix

The sum of the diagonal coefficients of the stiffness matrix is an interesting parameter to follow: this sum is called the Trace of the matrix (Tr(C_{ij})). The Trace is a global parameter independent of the study axis allowing an averaged view of the matrix: the study is not then made on six independent coefficients but on a single parameter.

Figure 5 presents the Trace of each sample from each individual according to their PMI and their mass densities. A color code is used to link the figure on the left and that on the right: the group G1 (1 – 3 years) is represented in blue, the group G2 (10 – 13 years) in red and groups G3-G4 (over 50 years later noted G3+4) in black. As for the diagonal coefficients of the matrix, two zones can be delimited for samples with high and low values of the Trace. The results show a margin of distinction of around 10 GPa, letting to separate samples with PMI lower and higher than 50 years (Figure 5A).

The precision of measurement of the diagonal coefficients has an impact on the Trace of the stiffness matrix. In the case of G1, the precision per individual is around 2.5% and two groups of blue points (between 80 and 90 GPa) can be distinguished from two groups of red points of G2 (between 95 and 105 GPa), where precision per individual is around 3.5%. Taking into account the high precision, the Trace representation can lead to possibility to distinguish groups G1 and G2. However, for PMI over 13 years, the precision per individual (and therefore per PMI group) is lower, and is not possible to assign a belonging of PMI group to a range of values of the Trace.

The Figure 5B presents the evolution of the Trace of stiffness matrix with respect to mass density. This representation let us again to define a delimitation zone whose margin of distinction is around 10 GPa. For the upper zone, the results show that for PMI lower than 50 years, the Trace is linearly related to mass density ($R^2 = 0.92$). Therefore, elastic and homogeneous hypotheses can be checked where mass density is predominant, since the Trace of the matrix is proportional to the mass density (common factor for all diagonal coefficients). However, for the lower zone, the results show that PMI over 50 years, the Trace follows less a linear behavior ($R^2 = 0.56$), for which a nonlinear trend would favor multi-parameter influence hypothesis (not only mass density).

As with the study of isolated diagonal coefficients, the evolution of the trace makes it possible to divide the groups of individuals into three distinct populations: this division is supported by the WMW test with a threshold of 0.01. The same test can be performed for density. The G1 // G2 and G2 // G3+4 groups can be divided into two populations (p-values < 0.002), this is not the case for the G1 // G3+4 groups whose p-value associated with density (p-value = 0.021), is distinct from that associated with trace (p-value < 0.001).

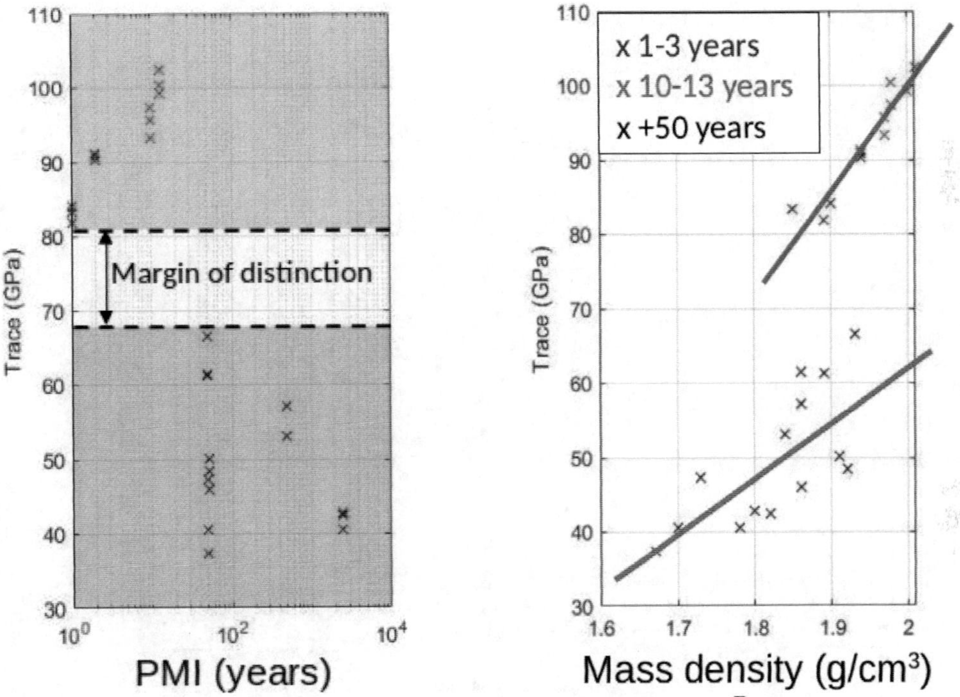

A B

Figure 5: Trace of stiffness matrix per individual as function of: A) PMI and B) Mass density.

Anisotropy ratios

The anisotropy ratios C_{11} / C_{22} and C_{44} / C_{55} were calculated for each sample from each individual and are shown in Figure 6. The average value of these ratios per PMI group as well as their standard deviations are listed in Table 4.

In general, the estimates of the ratios from the G3 and G4 groups are less repeatable than those of groups G1 and G2: the two youngest groups have standard deviations less than 6% while the two oldest groups have standard deviations included between 5% and 28%. These differences may be associated with the heterogeneity between individuals belonging to the same group and/or to differences between samples from the same individual.

The C_{11} / C_{22} and C_{44} / C_{55} ratios provide information on the anisotropy model applied to the bone. The average of the results per group shows that the bone is generally transversely isotropic to within 10%: the average ratios have values between 0.90 and 1.11. However this transverse isotropy hypothesis seems to evolve with the PMI: groups G1 and G2 have ratios closer to 1 (difference less than 0.03 for standard deviations less than 6%) than the ratios of groups G3 and G4 (difference greater than 0.07 for standard deviations between 5% and 28%). Considering these results, the post-mortem mechanical evolution of the bone would be stronger than the pre-mortem parameters between individuals. Therefore, elastic homogeneous transversely isotropic model could be used to highlight belonging of a bone to PMI lower than 50 years (even close to pre-mortem situation). However after 13 years, the post-mortem evolution (multi-parameter) makes it difficult to distinguish and classify bones according to their mechanical behavior. This last situation requires to take into account heterogeneity in fiber direction and another anisotropy model might be necessary.

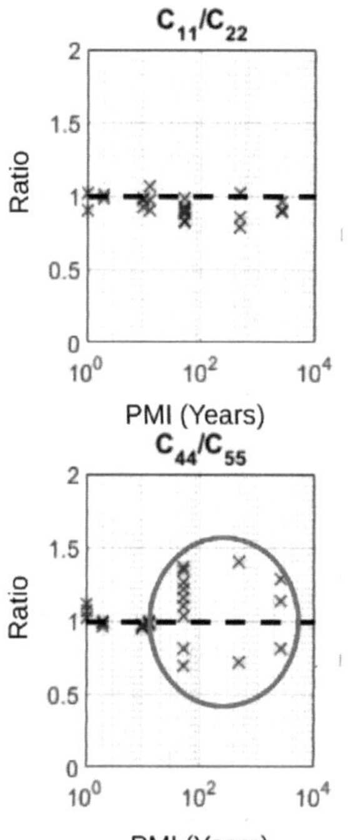

Figure 6: Measured anisotropy ratios for individuals as of PMI.

Table 4: Mean and standard deviation of anisotropy ratios as function of PMI group.

	C_{11} / C_{22}	C_{44} / C_{55}
G1: 1 – 3 years	0.97 ± 6%	1.02 ± 6%
G2: 10 – 13 years	0.97 ± 6%	0.97 ± 2%
G3: ~50 years	0.90 ± 5%	1.11 ± 21%
G4: Archeo	0.90 ± 9%	1.07 ± 28%

5. Conclusion

The technique presented here is a nondestructive experimental method for the characterization of post-mortem interval (PMI) of human bones. In order to validate the feasibility, an experimental technique was implemented based on test bench for the measurement of stiffness coefficients already used for bone ultrasonic assessment in medical applications. The study addressed the questions about the mechanical behavior of post-mortem bone and the relevance of elastic homogeneous transversely isotropic model and its validity according to PMI. For this, physical measurements and extraction of diagonal coefficients of the stiffness matrix were carried out. On one hand, the analysis of measurement precision and the use of the Trace let to validate the elastic and homogeneous hypotheses, where mass density is predominant. On the other hand, the study of anisotropy ratios let to study the behavior of this parameter during post-mortem alteration. Overall, the results of this work permitted to carry out a first classification of post-mortem bone relating physico-mechanical parameters and PMI. The agreement with literature between stiffness coefficients of short PMI close to pre-mortem condition allowed to validate the experimental approach.

In conclusion, this study contributes a new set of elasticity values for post-mortem human cortical bone, providing insights into the evolution of stiffness coefficients with post-mortem alteration. Moreover, it offers the first complete analysis of diagonal coefficients and the Trace of stiffness matrix, giving some indication of how bone anisotropy is related to time after death. Future perspectives may include studying the bone samples with PMI between 13 and 50 years and the use of complementary techniques. Wave attenuation might be a parameter of interest for further studies on scattering and viscoelasticity of post-mortem human bone, and techniques such as Resonant Ultrasonic Spectroscopy (which allows full stiffness matrix characterization) or opto-acoustic measurements (in order to reduce bone conditioning), could be considered.

Acknowledgment

This work has been funded by CY Cergy Université in collaboration with the IRCGN through the Post-MortUS project. The authors would like to thank Franck Nolot from IRCGN for his availability to the project and Laurent Sabatier from Laboratory of Mechanics and Acoustics of Marseille for the X-ray micro-tomography scans of the samples.

References

[1] G. Quatrehomme, *Traité d'anthropologie médico-légale*. De Boeck, 2015.
[2] S. Berg, "Determination of bone age", Interscience, 1962.
[3] J. Krautkrämer and H. Krautkrämer, *Ultrasonic testing of materials*. Springer Science & Business Media, 1990.
[4] P. Laugier and Q. Grimal, eds. *Bone quantitative ultrasound: new horizons*. Vol. 1364. Springer Nature, 2022.
[5] H.S. Yoon and J.L. Katz, "Ultrasonic wave propagation in human cortical bone—I. Theoretical considerations for hexagonal symmetry", Journal of Biomechanics, 9 (1976), pp. 407-412
[6] R.B. Ashman, S.C. Cowin, W.C. Van Buskirk, and J.C. Rice, "A continuous wave technique for the measurement of the elastic properties of cortical bone", Journal of Biomechanics, 17 (1984), pp. 349-361
[7] C. Baron, H. Follet, M. Pithioux, C. Payan, and P. Lasaygues, (2022), "Assessing the Elasticity

of Child Cortical Bone", In: Laugier, P., Grimal, Q. (eds) *Bone Quantitative Ultrasound*, Advances in Experimental Medicine and Biology, vol 1364. Springer, Cham. https://doi.org/10.1007/978-3-030-91979-5_14

[8] J. L. Rose, *Ultrasonic waves in solid media*. Cambridge University Press, 2014. https://doi.org/10.1017/CBO9781107273610

[9] R. B. Martin, D. B. Burr, N. A. Sharkey, and D. P. Fyhrie, *Skeletal Tissue Mechanics*. New York, NY: Springer New York, 2015. doi: 10.1007/978-1-4939-3002-9

[10] D. Royer and T Valier-Brasier. *Elastic Waves in Solids 1: Propagation*. John Wiley & Sons, Incorporated, 2022.

[11] P. Lasaygues and M. Pithioux, "Ultrasonic characterization of orthotropic elastic bovine bones", Ultrasonics, Volume 39, Issue 8, 2002, Pages 567-573, ISSN 0041-624X, https://doi.org/10.1016/S0041-624X(02)00261-5.

[12] X.N. Dong and X.E. Guo, "The dependence of transversely isotropic elasticity of human femoral cortical bone on porosity", J. Biomech., vol. 37, no. 8, pp. 1281–1287, Aug. 2004.

[13] E. Bossy, M. Talmant, F. Peyrin, L. Akrout, P. Cloetens, and P. Laugier, "An in vitro study of the ultrasonic axial transmission technique at the radius: 1-MHz velocity measurements are sensitive to both mineralization and intracortical porosity", J. Bone Miner. Res., vol. 19, no. 9, pp. 1548–1556, 2004.

[14] G. Haïat, S. Naili, Q. Grimal, M. Talmant, C. Desceliers, and C. Soize, "Influence of a gradient of material properties on ultrasonic wave propagation in cortical bone: Application to axial transmission", J. Acoust. Soc. Am., vol. 125, no. 6, pp. 4043–4052, Jun. 2009.

[15] E. Lefèvre, P. Lasaygues, C. Baron, C. Payan, F. Launay, H. Follet, and M. Pithioux, "Analyzing the anisotropic Hooke′s law for children′s cortical bone", Journal of the Mechanical Behavior of Biomedical Materials, Volume 49, 2015, Pages 370-377, ISSN 1751-6161, https://doi.org/10.1016/j.jmbbm.2015.05.013.

[16] B. Castagnède, J. T. Jenkins, W. Sachse, and S. Baste, "Optimal determination of the elastic constants of composite materials from ultrasonic wave-speed measurements", J. Appl. Phys., vol. 67, no. 6, pp. 2753–2761, Mar. 1990, https://doi.org/10.1063/1.345441

[17] D. Zellouf, Y. Jayet, N. Saint-Pierre, J. Tatibout, and J.C. Baboux, "Ultrasonic spectroscopy in polymeric materials. application of the Kramers-Kronig relations", J. Appl. Phys. **80**(5), 2728–2732 (1996).

[18] A. Arciniegas, H. Achdjian, J. Bustillo, F. Vander Meulen, and J. Fortineau, "Experimental Simultaneous Measurement of Ultrasonic Properties and Thickness for Defect Detection in Curved Polymer Samples", J Nondestruct Eval 36, 46 (2017). https://doi.org/10.1007/s10921-017-0427-3

[19] A. Arciniegas, L. Martinez, A. Briand, S. Prieto, S. Serfaty, and N. Wilkie-Chancellier, "Experimental ultrasonic characterization of polyester-based materials for cultural heritage applications", Ultrasonics, 2017, 81:127-134. doi:10.1016/j.ultras.2017.06.011

[20] B. K. Hoffmeister, S. R. Smith, S. M. Handley, and J. Y. Rho, "Anisotropy of Young's modulus of human tibial cortical bone", Med. Biol. Eng. Comput., vol. 38, no. 3, pp. 333–338, 2000.

Integrated analysis of materials for offshore wind turbine blades: mechanical and acoustical coupling

Khalid AOUJDAD[(1)], Elhadji-Amadou BA[(2)], Pierre MARECHAL[(1)], Damien LEDUC[(1)], Alexandre VIVET[(2)], Florian GEHRING[(2)], Mounsif ECH-CHERIF EL-KETTANI[(1)]

(1) Laboratory of Waves and Complex Media (LOMC CNRS UMR 6294), Le Havre Normandy University, Le Havre, France

(2) CIMAP, Normandie Univ, ENSICAEN, UNICAEN, CEA, CNRS, CIMAP, 14000 Caen, France

Khalid.aoujdad@doct.univ-lehavre.fr

Abstract

This work focuses on assessing the structural degradation of offshore wind turbine blade (OWTB) materials caused by exposure to the marine environment using acoustic and mechanical methods. Samples, consisting of a glass fibre reinforced polymer (GFRP) composite laminate with a styrene-acrylo-nitrile (SAN) foam core, are subjected to an accelerated hygrothermal ageing by immersion in seawater at 28-30% salinity, thermostated at 40°C. Non-destructive characterisation using ultrasonic waves and mechanical testing, including 3- and 4-point bending tests, are carried out. These results are in agreement with mechanical tests which show an 8% reduction in maximum stress for the same immersion time.

Keywords

Ultrasonic characterisation, glass fibre reinforced polymer (GFRP), composite material, accelerated seawater ageing, mechanical testing, offshore wind turbine blades (OWTB).

1. Introduction

Offshore wind energy plays a crucial role in the quest for sustainable power generation. This work is aimed at carrying out a comprehensive investigation of representative samples of offshore wind turbine blades (OWTB) composites. Various factors are conditioning mechanical properties and acoustic characteristics, such as environmental conditions, economic feasibility. OWTB are continuously improving in terms of power production performance and dimensions. They are exposed to high wind speeds driving high mechanical loads [1], and salt water exposure causing accelerated ageing. It is therefore essential to select materials that can withstand these conditions while maintaining structural integrity and preventing failure risks. The performance and reliability of wind turbine blades is largely determined by the mechanical properties. These include strength, stiffness, fatigue resistance, and durability. Composite materials, such as GFRP composites, are commonly used for wind turbine blades due to their high strength-to-weight ratio and fatigue resistance [2] [3]. However, the design of composite materials and manufacturing techniques should be based on comprehensive mechanical testing and analysis to ensure optimum performance and reliability.

This paper focuses on the combined analysis of materials for OWTB, with particular emphasis on the link between mechanical evaluation and ultrasonic NDT. Previous NDT works on wind turbine blades show that ultrasonic techniques are the most commonly used and have demonstrated their robustness and accuracy in detecting and diagnosing internal and external faults [4]. C-scan testing allows the detection of various flaws in wind turbine blades. It is suitable for quality control during manufacture and for assessing accumulated damage during fatigue testing. A portable C-scan scanner can also be used for in-situ inspections to assess damage during the normal life of the blade [5]. While [6] has shown through experimental work, that the Lamb guided wave technique is suitable for the ultrasonic (NDT) of wind turbine blades, particularly when access is limited to a single face.

Several factors influence the degradation of the blades of offshore wind turbine blades (OWTBs) compared to onshore wind turbines, with moisture and salinity of the water being critical parameters. Some studies have concluded that salt has no significant effect on the structural degradation of composites [7]. Studies on the effects of moisture highlight the critical importance of the interface in composites, particularly with regard to moisture-induced degradation [8]. They show that the initial strength of the system can influence the rate of degradation in the presence of moisture, highlighting the need to consider this interaction when designing strong composites. To simulate the marine environment, accelerated hygrothermal ageing is performed by immersing specimens in seawater at 30% salinity and heating to 40°C. This accelerated ageing process makes it possible to shorten the study period.

This integrated approach has great potential to advance our understanding of OWTB materials, leading to improved durability, reliability and performance in renewable energy systems. We aim to optimise material selection, design parameters and maintenance strategies by combining mechanical analysis and non-destructive ultrasonic testing. This paper presents a case-of-study analysis of ageing of OWTB in marine environment and includes three additional sections. In section two, the choice of materials is discussed on the basis of comparative research and analysis. The manufacturing process of the studied samples is presented. A description of the ageing process is then provided. The third and fourth sections are then devoted to the mechanical testing and acoustic characterisation of the specimens, respectively.

2. Material selection and sample fabrication

2.1. Sample selection

In the first phase of this work, a comparative analysis of the materials used to manufacture the blades was carried out based on a literature review [3]. The majority of studies have focused on composite materials due to their weight to performance ratio. Their performance includes resistance to various mechanical stresses and corrosion [9]. Other performance aspects are also considered, related to the wind turbine installation environment, such as the ability to withstand moisture and temperature fluctuations for offshore wind turbines [10].

In parallel with the literature review, a visit was made to LM Wind Power, a renowned global supplier that has been supplying blades to wind turbine manufacturers worldwide for over 40 years and has extensive expertise in this field. As a result, our prototype manufacturing process started on the basis of LM Wind Power blade fabrication plant in Cherbourg, France. The sandwich composite samples (Figure 1) are consisting of two skins of unidirectional glass fibre reinforced polyester (GFRP) composites wrapping a styrene-acrylo-nitrile (SAN) foam core.

Figure 1. Sandwich composite plate: three layered GFRP/Foam/GFRP.

2.2. Manufacturing process

As illustrated in Figure 2, a composite material was developed by mixing glass fibres, polyester resin, and a Styrene-AcryloNitrile (SAN) foam core.

Polyester resin Synolite 3720-I-1 One-directional glass-fibre roll SAN foam core ⌀10mm

Figure 2. Composite components: polyester resin, glass fibres, and SAN foam core materials.

The manufacturing process (Figure 3) involved the impregnation of the glass fibres (*Sicomin*™ E-glass L500 127cm) with an isophthalic based polyester resin (*Samaro*® SYNOLITE™ 3720-L-1). A vacuum pressure of is applied to ensure uniform distribution and optimal bonding between the components. The vacuum infusion moulding process enhances the consolidation of the composite material, resulting in improved strength and integrity.

Impregnation process | Assembly of the mould | Vacuum pressure application

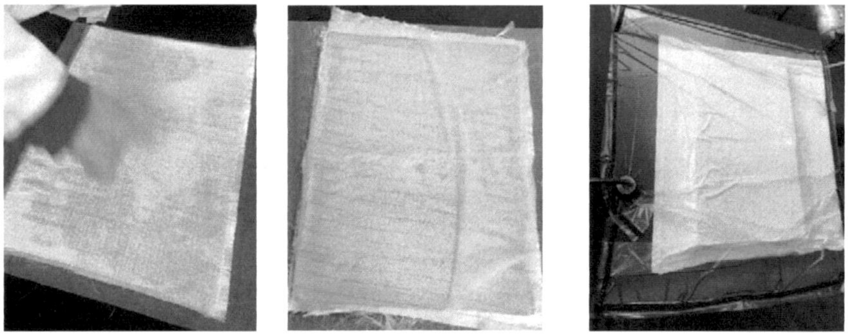

Figure 3. Samples manufacturing process.

2.3. Accelerated ageing

Since marine environment is a challenging operating environment, a rigorous assessment of material behaviour is essential prior to implementation on site. By using a temperature accelerated ageing process, the harsh environmental conditions are simulated, as experienced by OWTB, enabling a comprehensive study of material degradation over the extended service life of OWTB [11]. The accelerated ageing of the fabricated sandwich composite samples is processed by immersion in a tank filled with seawater regulated at a temperature of 40°C (Figure 4). This process, standing for an accelerated hygrothermal ageing, is designed to assess the long-term durability and performance of composites in a relatively short time frame.

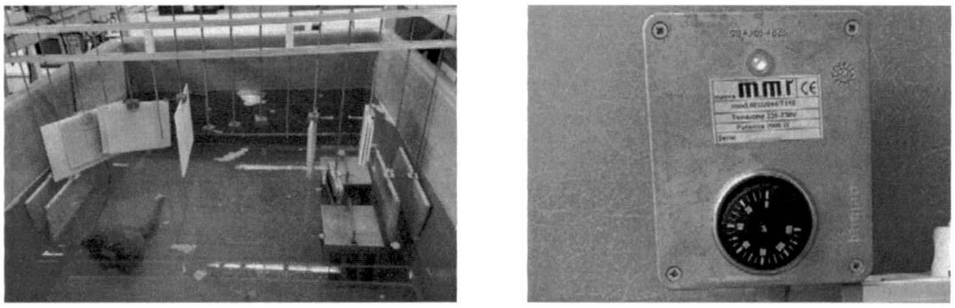

Figure 4. Tank filled with seawater and thermostated at 40°C to accelerate the aging process.

3. Determination of mechanical properties after ageing

The mechanical analysis aims to investigate the residual mechanical properties and the kinetics of internal damage growth after ageing in composite materials for wind turbine blade applications. The experimental mechanical testing (Figure 5) includes instrumented flexure tests, which provide an overview of mechanical behaviour such as bending strength and Young's modulus. In-situ monitoring of acoustic emissions allows the detection of internal damage mechanisms to be detected in real time. Visualisation of microstructural changes can be achieved by post-mortem microstructural investigations. In addition, optical microscopy and C-scan mapping are used to improve our understanding by providing detailed visualisation and analysis of microstructural features and internal damage at different length scales.

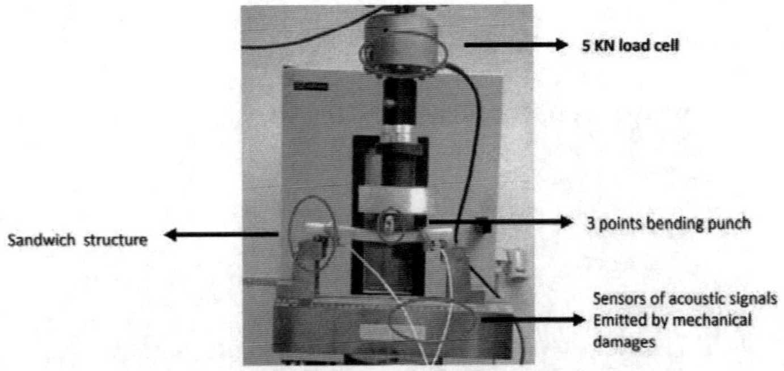

Figure 5. Bending test machine instrumented with acoustic emission sensors.

These mechanical tests identified the acoustic signature of 5 types of internal damage. Further work consists of monitoring the evolution of the acoustic signature with ageing of the 5 types of damage identified, i.e. matrix cracks, fibre/matrix decohesion, skin/core decohesion, delamination and fibre breakage [12]. As illustrated by Figure 6 a decrease by 8% and 18% in maximal stress is evaluated after ageing periods of 16 and 30 weeks, respectively.

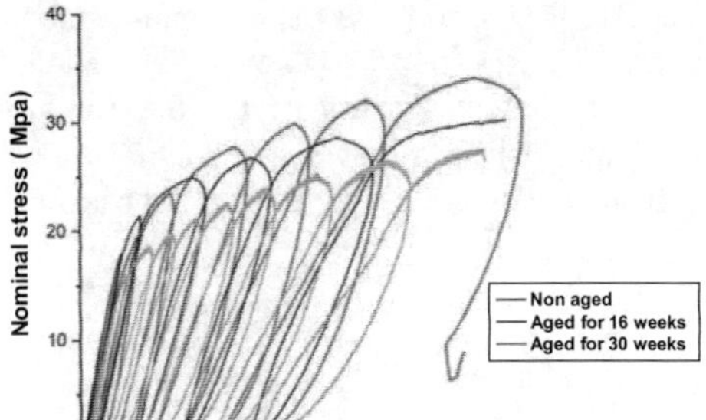

Figure 6. Nominal stress evolution versus axial displacement of RPL 4-point bending sandwich structures.

Journal of Physics: Conference Series **2904** (2024) 012004 doi:10.1088/1742-6596/2904/1/012004

4. Non-destructive ultrasonic characterisation

An experimental setup (Figure 7) operating in pulse/echo mode enables bulk waves to be propagated through the depth of the composite material and give information on the material. The sample is immersed in a thermostatically controlled tank and a planar transducer emits ultrasonic waves which propagate through the thickness of the sample. An Olympus V320 focusing immersion transducer is used, with a centre frequency of 7.5 MHz and a diameter of 1/2 inch. This experiment is carried out in the pulse mode, and the same transducer records the echoes reflected at the various interfaces. The chosen indicator to monitor the sample's state of health is the longitudinal velocity. In addition, the *Mistras* device not only measures velocity, but also provides surface C-scan images through the thickness, providing slice-by-slice visual representation of the sample's internal and potential anomalies. This capability is critical when examining aged samples, as it provides insight into structural changes and potential defects that may have occurred over time as a result of ageing processes.

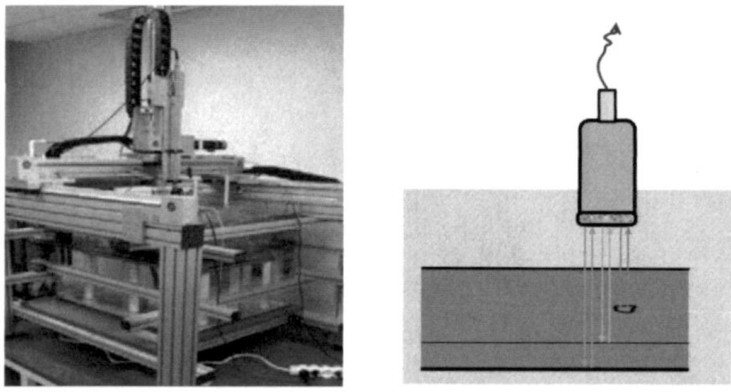

Figure 7. Acoustic experimental setup (*Mistras* device).

4.1. C-scan mapping of the skin sample

In this study, C-scan mapping is used to inspect skin samples and provide insight into their internal structure. The 4 plies samples are 2.1 ± 0.3 mm thick, and the 6 plies samples are 4.1 ± 0.3 mm thick. The C-scan performed was carried out on the echo from the GFRP skin, using three different time gates (Figure 8), each spanning 485 ns. Those gates correspond to three sections of the sample (Figure 9): the surface, the internal structure and the bottom interface.

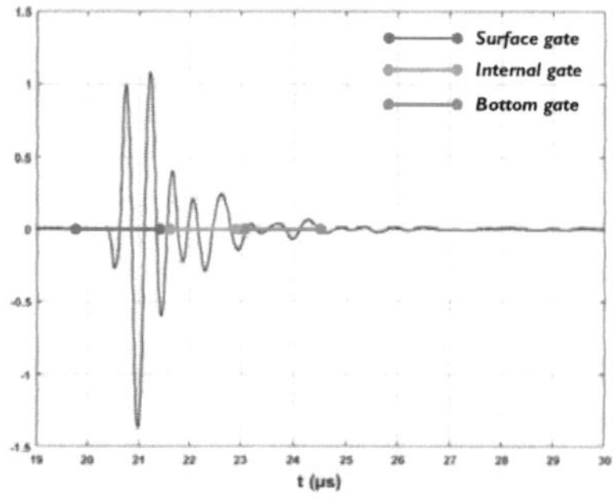

Figure 8. Time gates of the C-scan.

Changes before and after the ageing process are illustrated in Figure 9. On the surface C-scan only the pattern of weft threads (vertical red lines) can be seen before ageing, because this top layer is covered by a light layer of resin. After 28 weeks, this latter is damaged, making the pattern of fibres in the 0° direction observable. The background C-scan shows a strong attenuation and makes it difficult to assess the bottom, indicative of a possible transition to a viscoelastic state and behaviour [13].

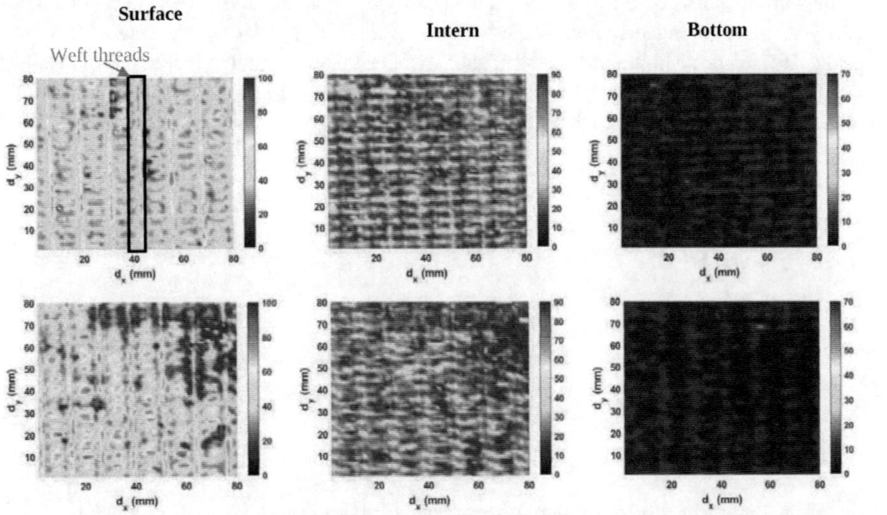

Figure 9. C-scans of GFRP 6 plies plate unaged (top) and aged for 28 weeks at 40°C (bottom).

4.2. Changes in longitudinal velocity with ageing

A velocity measurement is performed with a *PVA Tepla* SAM 301 acoustic microscope (Figure 10). An immersion transducer with a centre frequency of 7.5 MHz and a half inch diameter is used to ensure accurate longitudinal wave velocity measurements.

Figure 10. *PVA Tepla* SAM 301 acoustic microscope.

The measurement is performed on a 4 plies skin sample, 2.1 ± 0.3 mm thick, aged for 15 weeks. The experimental settings (temperature and acquisition settings) were carefully fixed to ensure the reproducibility of the measurement in all tests. This meticulous control of the experimental conditions was essential to minimise measurement error, (about ± 5%) that is primarily influenced by sample thickness variation and surface uniformity. As illustrated by Figure 11, a 134 ns difference in the time-of-flight of the echo reflected from the bottom of the aged sample. With a skin thickness of 2.1 mm, the wave velocity of the unaged sample is measured to be 2736 m/s, while the aged sample has a velocity of 2516 m/s. relative decrease of 8% after ageing. This decrease highlights the structural changes that occur in the material as a result of the ageing process. This underlines the interest of monitoring wave velocity as a sensitive indicator of skin degradation.

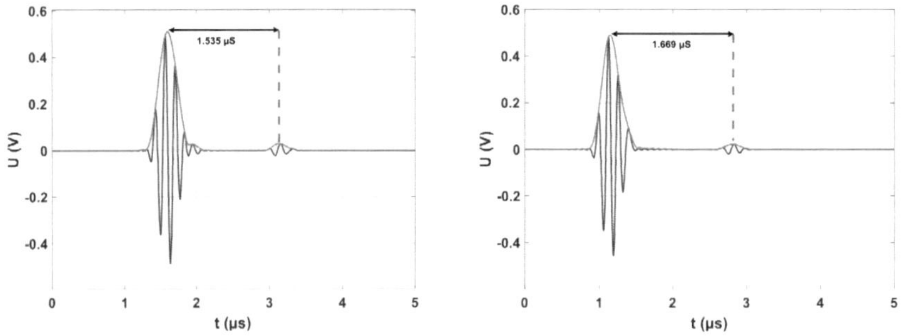

Figure 11. Reflected signals on unaged (left) and aged (right) samples.

5. Conclusion

The mechanical tests conducted have demonstrated that maximum stress serves as a reliable indicator of structural degradation of GFRP composite materials resulting from hydrothermal ageing. Similarly, longitudinal velocity emerges as a suitable acoustic indicator. The consistent evolution of these two indicators, both mechanically and acoustically, underscores their reliability in gauging degradation, and their proportional relationship reinforces their effectiveness. A comprehensive study, integrating simultaneous mechanical and acoustic measurements, is underway to further explore these findings.

Acknowledgments

The authors acknowledge the financial support of the French Agence Nationale de la Recherche (ANR) LabEx EMC3 through the project COSPHI (Grant No. 10-LABX-0009), and the Normandy Region (Réseau d'Intérêt Normand - Label d'Excellence).

References

[1] J.-T. Horn et B. J. Leira, « Fatigue reliability assessment of offshore wind turbines with stochastic availability », *Reliability Engineering & System Safety*, vol. 191, p. 106550, 2019, doi: 10.1016/j.ress.2019.106550.

[2] P. Brøndsted, H. Lilholt, et A. Lystrup, « Composite Materials for Wind Power Turbine Blades », *Annual Review of Materials Research*, vol. 35, n° 1, p. 505-538, 2005, doi: 10.1146/annurev.matsci.35.100303.110641.

[3] L. Thomas et M. Ramachandra, « Advanced materials for wind turbine blade- A Review », *Materials Today: Proceedings*, vol. 5, n° 1, p. 2635-2640, 2018, doi: 10.1016/j.matpr.2018.01.043.

[4] F. P. García Márquez et A. M. Peco Chacón, « A review of non-destructive testing on wind turbines blades », *Renewable Energy*, vol. 161, p. 998-1010, 2020, doi: 10.1016/j.renene.2020.07.145.

[5] « Nondestructive evaluation (NDE) of composite/metal bond interface of a wind turbine blade uskng an acousto-ultrasonic technique », Aerospace Sciences Meetings. Consulté le: 8 avril 2024. [En ligne]. Disponible sur: https://arc.aiaa.org/doi/10.2514/6.1997-959

[6] R. Raišutis, E. Jasiūnienė, et E. Žukauskas, « Ultrasonic NDT of wind turbine blades using guided waves », *Ultragarsas / Ultrasound*, vol. 63, n° 1, Art. n° 1, 2008.

[7] K. Liao, C. R. Schultheisz, et D. L. Hunston, « Effects of environmental aging on the properties of pultruded GFRP », *Composites Part B: Engineering*, vol. 30, n° 5, p. 485-493, 1999, doi: 10.1016/S1359-8368(99)00013-X.

[8] S. S. Wang, S. S. Wang, et D. W. Fitting, « Composite materials for offshore operations », National Institute of Standards and Technology, Gaithersburg, MD, NIST SP 887, 1995. doi: 10.6028/NIST.SP.887.

[9] A. A. Shittu, A. Mehmanparast, M. Shafiee, A. Kolios, P. Hart, et K. Pilario, « Structural reliability assessment of offshore wind turbine support structures subjected to pitting corrosion-fatigue: A damage tolerance modelling approach », *Wind Energy*, vol. 23, n° 11, p. 2004-2026, 2020, doi: 10.1002/we.2542.

[10] M. Nachtane, M. Tarfaoui, D. Saifaoui, et K. Hilmi, « Hygrothermal and mechanical performance evaluation of glass-polyester composite for renewable marine energies », in *13ème congrès de mécanique (cmm2017)*, Meknès, Morocco, 2017. Consulté le: 11 janvier 2024. [En ligne]. Disponible sur: https://hal.science/hal-01728047

Journal of Physics: Conference Series **2904** (2024) 012005 doi:10.1088/1742-6596/2904/1/012005

Multi-modal characterization of ultrasonic bulk wave properties in heterogeneous textured media through finite element computations

Vincent Dorval*, **Nicolas Leymarie, Alexandre Imperiale and Edouard Demaldent**

Université Paris-Saclay, CEA, List, F-91120, Palaiseau, France

*E-mail: vincent.dorval@cea.fr

Abstract. Ultrasonic Non Destructive Testing can be heavily affected by heterogeneous materials, such as some coarse-grained metals, concretes or composites materials. Understanding and predicting the behaviour of ultrasonic waves in such complex media has been a longstanding focus of research. Originally, theoretical analytical models have been developed. Their utility tends to be constrained by their inherent approximations and by the difficulties associated with verifying them. In recent years, transient Finite Element computations have been used to overcome these limitations. By simulating the propagation of ultrasonic waves through representative volumes of the material, it is possible to determine the properties of the coherent wave. However, this approach has been mainly applied to longitudinal waves only, and the analysis methods used are not necessarily applicable to shear waves. This communications shows how, with a dedicated simulation setup, it is possible to characterize the behaviour of both longitudinal and shear waves. Both waves can be characterized using a single computation, saving on computation times. The presented methodology is also applicable to statistically anisotropic media, such as textured metallic microstructures, in which three distinct modes can be observed in the coherent field: one quasi-longitudinal and two quasi-shear modes. These three modes can be characterized in a single simulation. One of the objectives of this technique is to provide material input data for macroscopic scale simulations using ray-based or high order Finite Element models in application to the control of various metallic parts, including welded components with textured regions.

Introduction

Finite Element simulations in representative volumes of a microstructure has emerged as a means of characterizing the behaviour of ultrasonic bulk waves in heterogeneous materials [1], [2], [3]. Previous works have mainly focused on longitudinal waves. With an appropriate simulation setup, shear waves can also be investigated. This communication demonstrates how both modes can be included in the same simulation, and how this approach extends to statistically anisotropic materials where three different modes can exist.

The first section presents the general principles of the method that uses Finite Element computations in representative elementary volumes of a microstructure to estimate phase velocity and attenuation. Hill averages of the material, which are used both for results comparison

Journal of Physics: Conference Series **2904** (2024) 012005 doi:10.1088/1742-6596/2904/1/012005

and as an input of the method, are detailed in the second section. The third and fourth section present examples of applications of the method to statistically isotropic and anisotropic materials.

1. Using Finite Element computations to estimate phase velocities and attenuations in heterogeneous materials

The basic principles of the approach used in the present work are similar to those of the works of Van Pamel [1] or Ryzy [2]. Random Voronoi tessellations are used to generate virtual samples of polycrystalline media. The propagation of ultrasonic plane waves through these samples is computed using Finite Element Methods. The evolution of these plane waves as a function of depth is analysed in order to estimate the effective complex wavenumber of the coherent wavefront and thus deduce the attenuation and phase velocity as a function of frequency. The computing tools used for this study are implemented in a development version of the CIVA platform [4].

A distinctive feature of these tools is their suitability for both L and S waves. To ensure it, they use periodic Voronoi tessellations as well as periodic boundary conditions for the FEM computation. Periodic boundary conditions also have other advantages over the often-used symmetric boundary conditions: even when S waves are not considered, symmetric boundary conditions effectively mirror grains at the boundary, which skews volume distribution [3]. An illustration of differences between symmetric and periodic simulation domains is shown in Figure 1. Periodicity applies to both the Voronoi tesselations and the FEM boundary conditions.

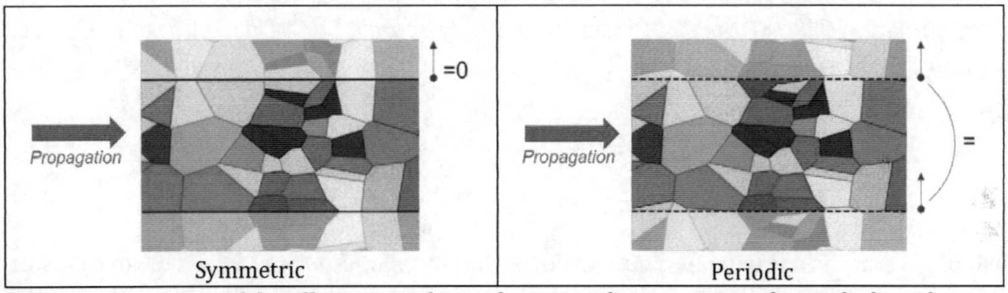

Figure 1. Illustration of the effective simulation domain with symmetric and periodic boundary conditions.

The FEM computation used here relies on a low-order scheme on a regular grid.

Plane waves are analysed through their average across a series of planes at several propagation distances, in a manner similar to one that was used in studies of concrete [5]. This yields an estimation of attenuation and phase velocity over a range of frequency for the considered wave. This approach can be applied to simulations where plane L and S waves have been generated simultaneously, as detailed in a subsequent section. An extension to the case of statistically anisotropic media where three modes can propagate is presented afterwards. The simultaneous analysis of several modes can reduce computation times in the isotropic case, while also providing a way to address mode selection problems in the anisotropic case.

2. Hill averages

The adaptation of the method to the anisotropic case presented in this communication requires an initial estimate of the effective elastic constants of the homogenised medium. The Hill average of elastic constants has been shown to provide a close agreement to measurements of effective properties of crystalline aggregates [6]. It is used here for comparisons with FEM results, and also as an initial estimate required by the post-processing method in the anisotropic case.

The Hill average is defined as the average of Voigt and Reuss averages. The Voigt average [7] correspond to an average of a heterogeneous material under an assumption of uniform strain. The Reuss average [8] corresponds to an assumption of uniform stress. In practical computations, these two averages are obtained by spatially averaging elastic constants for the Voigt average, and the compliance constants (inverse of the elastic constants) for the Reuss average.

2.1 Averages for a statistically isotropic medium

The statistically isotropic medium considered here has uniformly distributed grain orientations. The Voigt averaged medium is obtained by averaging the elastic constants over possible orientations. For a statistically isotropic medium, with no preferential orientation, this average can be expressed as follows:

$$\boldsymbol{C}^{isoV} = \frac{1}{8\pi^2}\int_{\varphi_1=0}^{2\pi}\int_{\Phi=0}^{\pi}\int_{\varphi_2=0}^{2\pi} \boldsymbol{R}(\boldsymbol{C},\varphi_1,\Phi,\varphi_2)sin(\Phi)\mathrm{d}\varphi_1\mathrm{d}\Phi\mathrm{d}\varphi_2. \tag{1}$$

$\boldsymbol{R}(\boldsymbol{C},\varphi_1,\Phi,\varphi_2)$ denotes the rotation of the single crystal elastic tensor $\boldsymbol{C}$ for the Euler-Bunge angles $(\varphi_1,\Phi,\varphi_2)$. The result of this average is an isotropic elastic tensor. Isotropic elastic tensors verify the relation $C_{12}^{iso} = C_{11}^{iso} - 2C_{44}^{iso}$ and are entirely defined by C_{11}^{iso} and C_{44}^{iso}. In the case of the Voigt average for single crystal elastic tensors with cubic symmetry, the following values are obtained:

$$C_{11}^{\mathrm{iso}V} = \frac{3C_{11} + 2C_{12} + 4C_{44}}{5}, \quad C_{44}^{\mathrm{iso}V} = \frac{C_{11} - C_{12} + 3C_{44}}{5}. \tag{2}$$

The Reuss average is obtained by averaging over the compliance tensor $\boldsymbol{S}$ (inverse of $\boldsymbol{C}$) instead of the elastic tensor $\boldsymbol{C}$ in equation (1). The averaging of $\boldsymbol{S}$ over orientations yields the Reuss-averaged compliance, that can then be inversed to produce the Reuss-averaged elasticity. For single crystal elastic tensor with cubic symmetry, the following results are obtained:

$$C_{11}^{isoR} = \frac{{C_{11}}^2 + C_{11}C_{12} + 8C_{11}C_{44} - 2{C_{12}}^2 - 4C_{12}C_{44}}{3C_{11} - 3C_{12} + 4C_{44}}, \quad C_{44}^{\mathrm{iso}R} = \frac{5(C_{11} - C_{12})C_{44}}{3C_{11} - 3C_{12} + 4C_{44}}. \tag{3}$$

The Hill-averaged elastic tensor is simply the average of the Voigt and Reuss averaged and is given by:

$$C_{11}^{\mathrm{iso}H} = \frac{C_{11}^{\mathrm{iso}V} + C_{11}^{\mathrm{iso}R}}{2}, C_{44}^{\mathrm{iso}H} = \frac{C_{44}^{\mathrm{iso}V} + C_{44}^{\mathrm{iso}R}}{2}. \tag{4}$$

The corresponding velocities for L and S waves can be obtained based on the averaged elastic constants and on the mass density ρ:

$$v_{LH} = \sqrt{\frac{C_{11}^{\mathrm{isoH}}}{\rho}}, v_{SH} = \sqrt{\frac{C_{44}^{\mathrm{isoH}}}{\rho}}. \tag{5}$$

2.2 Averages for a fibre texture

Welds, among other metallic parts, are known to behave as anisotropic media at the scale of ultrasonic inspection. This anisotropy can be related to preferential crystallographic orientations that stem from the welding process and create fibre-like textures [9]. A simplified representation of a weld is considered here, with elongated grains and a pure fibre texture along the elongation axis. This texture forces the alignment of one axis of the crystallographic frame with the fibre axis.

The fibre texture is aligned on third axis of the coordinate system. In this framework, crystallographic orientations are entirely defined by the first Euler-Bunge angle. The Voigt-averaged elasticity for the textured material can be computed as follows:

$$\boldsymbol{C}^{\boldsymbol{texV}} = \frac{1}{2\pi}\int_0^{2\pi} \boldsymbol{R}(\boldsymbol{C},\varphi,0,0)\mathrm{d}\varphi. \tag{6}$$

The results are transversely isotropic. The following results are obtained for the textured Voigt average texV:

$$\begin{gathered} C_{11}^{texV} = C_{22}^{texV} = \frac{3C_{11}+C_{12}+2C_{44}}{4}, \\ C_{33}^{\mathrm{tex}V} = C_{11}, \\ C_{12}^{\mathrm{tex}V} = \frac{C_{11}+3C_{12}-2C_{44}}{4}, \\ C_{13}^{\mathrm{tex}V} = C_{23}^{\mathrm{tex}V} = C_{12}, \\ C_{44}^{\mathrm{tex}V} = C_{44}^{\mathrm{tex}V} = C_{44}, \\ C_{66}^{\mathrm{tex}V} = \frac{C_{11}^{texV}-C_{12}^{texV}}{2}. \end{gathered} \tag{7}$$

The Reuss average is obtained by averaging compliance instead of elasticity, and also produces transverse istropic elasticity:

$$\begin{gathered} C_{11}^{texV} = C_{22}^{texV} = \frac{3C_{11}+C_{12}+2C_{44}}{4}, \\ C_{33}^{\mathrm{tex}R} = C_{11}, \\ C_{12}^{\mathrm{tex}R} = \frac{C_{11}{}^2-2C_{11}C_{44}-C_{12}{}^2+6C_{12}C_{44}}{2C_{11}-2C_{12}+4C_{44}}, \\ C_{13}^{\mathrm{tex}R} = C_{23}^{\mathrm{tex}R} = C_{12}, \\ C_{44}^{\mathrm{tex}R} = C_{44}^{\mathrm{tex}R} = C_{44}, \\ C_{66}^{\mathrm{tex}R} = \frac{C_{11}^{texR}-C_{12}^{texR}}{2}. \end{gathered} \tag{8}$$

Journal of Physics: Conference Series **2904** (2024) 012005 doi:10.1088/1742-6596/2904/1/012005

3. Application to statistically isotropic polycrystals

Virtual samples with dimensions 1×1×3 mm³ are considered. Waves propagate along the 3 mm length. The emitted waves are Ricker pulses with a central frequency of 15 MHz. They are produced by applying a uniform stress on one face of the sample. The stress is along the propagation direction for L waves, normal to it for S waves, or a mixture of both. The density and single crystal elastic constants of the grains were set to imitate those of 316L steel, and are given in Table I.

The sample contains 1000 grains per cubic millimetre, obtained by a Voronoi tessellation from uniformly distributed random nuclei. The elastic properties of each grain are determined by assigning them an orientation based on a uniform distribution, and rotating single crystal elastic constants according to that orientation.

The FEM grid step in these computations is set to 10 μm, which corresponds to approximately 0.05 times the expected wavelength for S waves in isotropic steel at 15 MHz, and to 0.1 times a typical grain dimension defined as the cubic root of the average grain volume. With these parameters, one computation takes approximately 5 minutes on a laptop computer with an eight-core Intel i9 processor and a 32 Go RAM.

Table 1. Single crystal elastic constants of austenite used to define the material properties of 316L steel microstructures [10].

ρ (g. cm-3)	C_{11} (GPa)	C_{12} (GPa)	C_{44} (GPa)
7.958	207	133	117

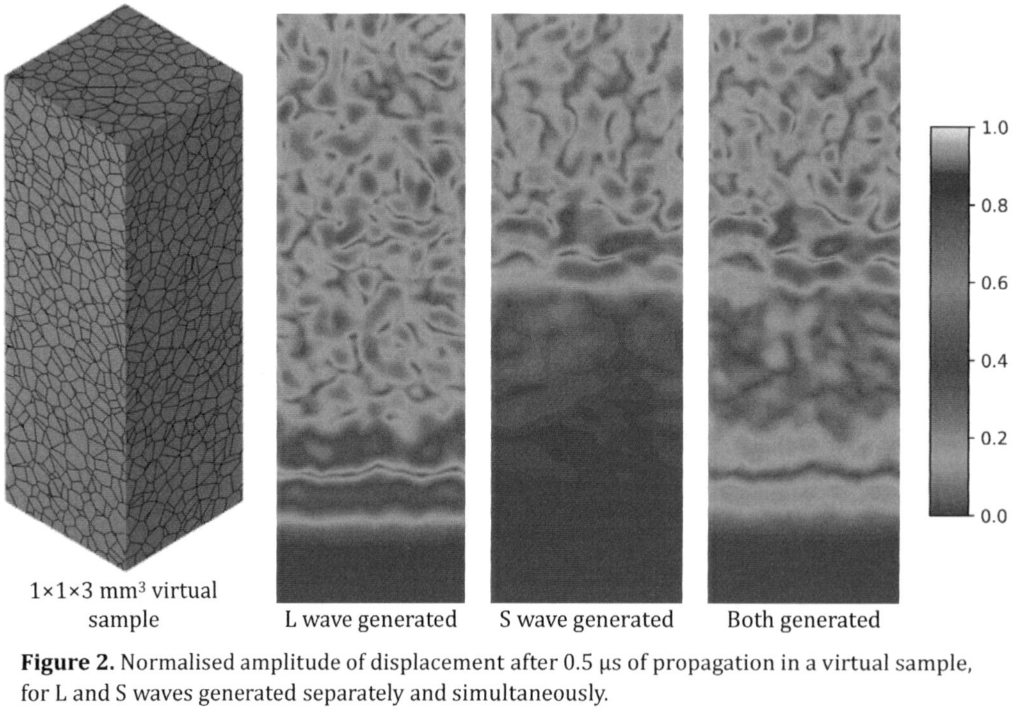

Figure 2. Normalised amplitude of displacement after 0.5 μs of propagation in a virtual sample, for L and S waves generated separately and simultaneously.

An example of a random sample and the ultrasonic fields generated for the three types of applied stress is shown in Figure 2.

L and S waves generated simultaneously are mixed in the representation of the norm of the displacement of Figure 2. When the vector components of the displacement along different axis are considered, as in Figure 3, they are clearly separated. The averages over x and y as a function of z that are plotted here are the same as those used in the post-processing method. Normalisation in this figure as well as in Figure 6 and Figure 7 is relative to the maximum displacement vector amplitude.

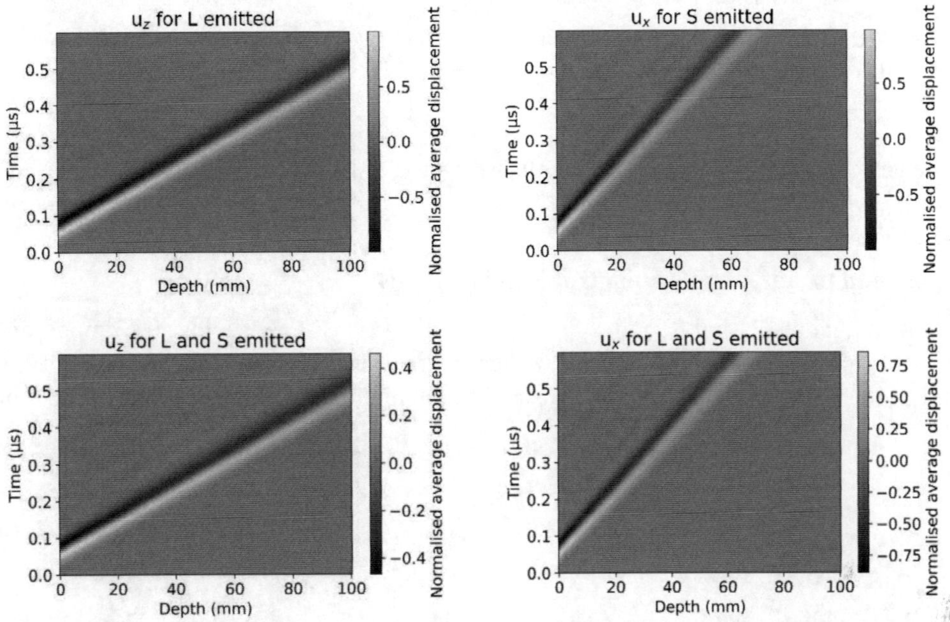

Figure 3. Main components of displacement averaged over x and y for the different waves considered.

The displacements projected along the relevant directions for L and S modes in the simultaneous computation are very similar to those obtained in two separate computations.

The displacements corresponding to each wave are processed in the frequency domain to obtain attenuations and velocities. Attenuation is related to the decrease of the amplitude, and velocity to the phase shift. The results obtained for both L and S waves, generated separately or simultaneously and averaged over 200 random microstructure samples, are plotted Figure 4. Confidence intervals are obtained based on the standard deviations of the 200 results.

The results obtained for simultaneous generation of L and S waves are in agreement with those obtained for separate generation. This constitutes an interesting way to save on computation times, since the simultaneous generation only requires half as many computations.

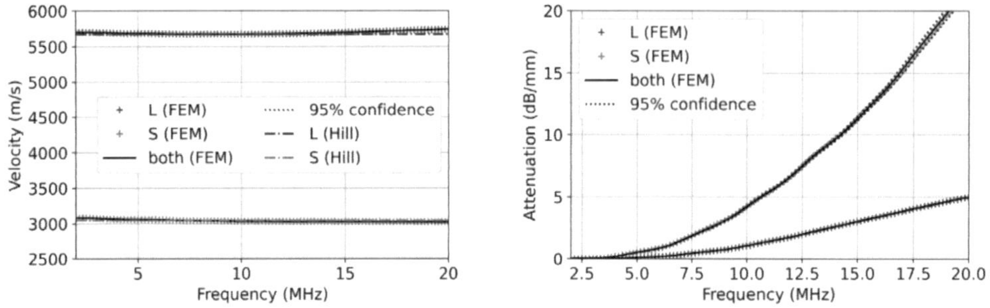

Figure 4. Estimated velocity and attenuation as a function of frequency averaged over 200 simulations, obtained for two modes computed separately and simultaneously.

The velocities corresponding to the Hill averages are also shown in Figure 4 and are in close agreement with numerical results.

4. Application to statistically anisotropic polycrystals

In anisotropic media, L and S modes are replaced by a quasi-longitudinal (qL) mode and two quasi-shear (qS1 and qS2) modes. The velocities and polarization directions of these three modes depend on their propagation directions. An adaptation of the post-processing method is required,

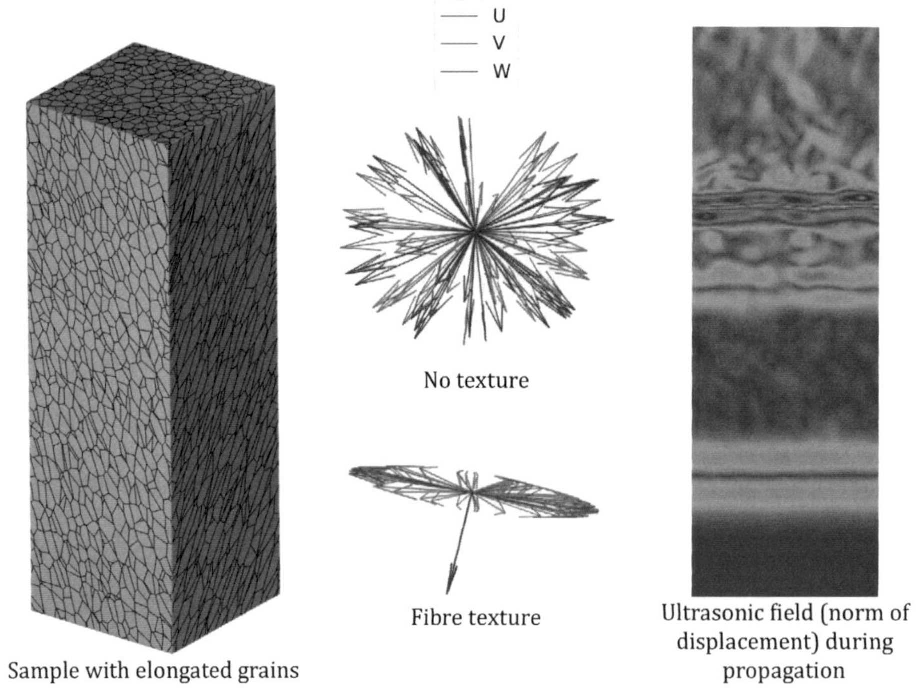

Figure 5. Elongated polycrystalline sample, illustration of crystallographic orientations without texture and with a fibre texture, and ultrasonic field computed in a sample with fibre.

in order to address the non-trivial polarizations and the coexistence of two shear modes with different velocities.

Statistically anisotropic samples are generated using elongated Voronoi tessellations and fibre-textured grain orientations. The elongation is obtained by computing the Voronoi tessellation in a space distorted by a factor 5 along that direction. The texture is obtained by aligning the third crystallographic axis with that elongation direction. The axis of the crystallographic frame are noted $(\boldsymbol{u},\boldsymbol{v},\boldsymbol{w})$ here, to avoid confusion with the frame $(\boldsymbol{x},\boldsymbol{y},\boldsymbol{z})$ of the sample. The elongation and texture axis is along the direction defined by a vector [1,2,3] in the sample referential. That particular direction is only chosen as an example to illustrate the principles of the method, which are applicable to any other directions.

Stress is applied in the [1,1,1] direction, with aim of generating all three modes. Three distinct waves corresponding to the three expected modes appear. The microstructure and the computed field are illustrated Figure 5.

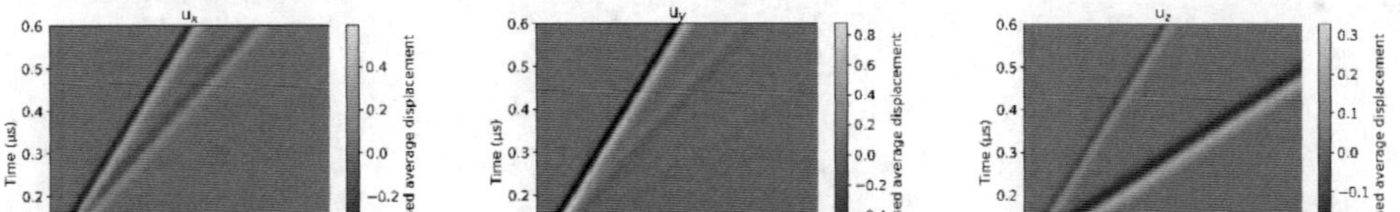

Figure 6. Displacement in the x, y and z direction averaged over x and y, in the anisotropic case.

Figure 6 shows, that unlike in the isotropic case, the modes are not separated along the directions of the frame of the sample.

This result is expected for an anisotropic material. The polarisation of a qL mode is usually not aligned with the propagation directions, and those of the qS1 and qS2 are usually not perpendicular to it. In order to try to predict the actual polarisation of these modes, the Hill average of the material is considered. By solving the Christoffel equation for the Hill average of the fibre-textured material with the considered orientation, the following polarisations are obtained for the three modes in the sample frame:

$$\boldsymbol{p}_{qS1} \approx \begin{bmatrix} -0.894 \\ 0.447 \\ 0.000 \end{bmatrix}, \boldsymbol{p}_{qS2} \approx \begin{bmatrix} -0.444 \\ -0.889 \\ -0.113 \end{bmatrix}, \boldsymbol{p}_{qSL} \approx \begin{bmatrix} -0.051 \\ -0.101 \\ 0.993 \end{bmatrix}. \tag{6}$$

The displacements projected over these three polarisations are plotted Figure 7.

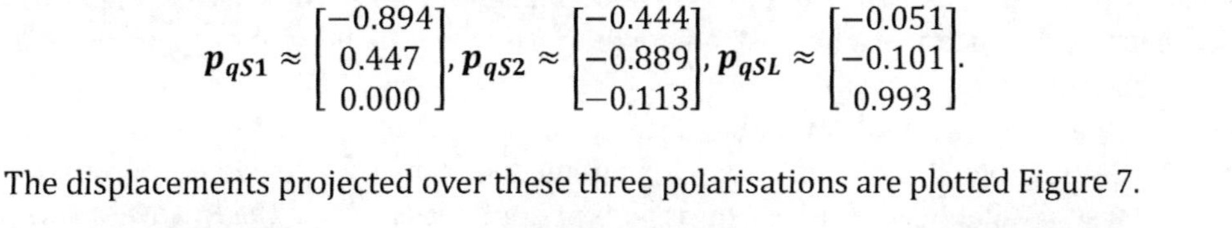

Figure 7. Displacement projected along the expected polarisations for qS1, qS2 and qL, averaged over x and y, in the anisotropic case.

After projection over these three polarisations, the three modes are clearly separated. The Hill average proves to be a useful approximaton of the effective medium. These projections are then projected as separate modes in order to estimate their velocities and attenuations. Results are plotted Figure 8.

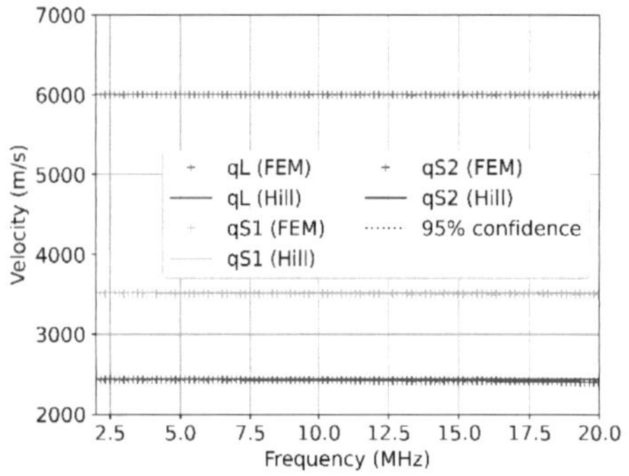

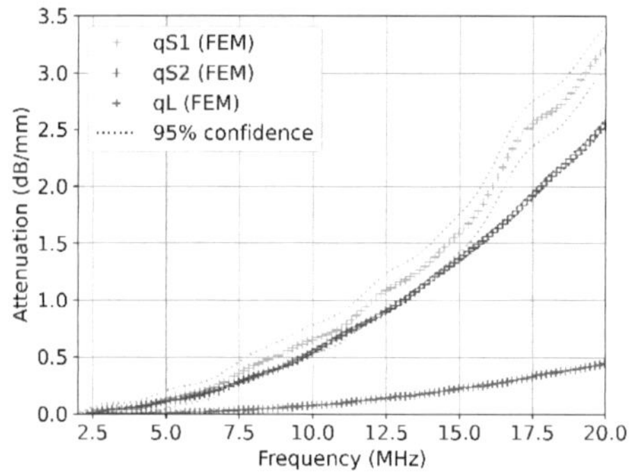

Figure 8. Estimated velocity and attenuation as a function of frequency averaged over 200 simulations, obtained for the displacements projected over the three polarisations corresponding to the Hill-averaged material.

The obtained velocities are again consistent with the predictions from the Hill average. The evaluation of attenuation for the qS1 mode shows significant uncertainty compared to the other modes. This could be related to the fact that it has lower amplitude than qS2, as shown in Figure 7, and imperfectly separated, making it sensitive to perturbations arising from it. Applying a load that excites the three modes more equally could help produce more balanced confidence intervals.

Conclusion

The results presented in this communication illustrates how the properties of several modes can be simultaneously obtained using Finite Element simulations of plane waves propagating through polycrystalline samples. The propagation of multiple modes is made possible by the use of periodicity for both microstructures and FEM boundary conditions. The possible applications of this approach include statistically anistropic materials, which include fiber-textured materials similar to those observed in welds.

In the anisotropic case, an estimation of the properties of the effective material is obtained using a Hill average. In the present case, a simplified texture was considered and analytical expressions were available. Analytical expressions are not necessary, as the Hill average can also be evaluated numerically by averaging over the microstructure description. The approach is therefore applicable to a wide range of isotropic and anisotropic materials.

References

[1] A. Van Pamel, C. R. Brett, P. Huthwaite, and M. J. S. Lowe, 'Finite element modelling of elastic wave scattering within a polycrystalline material in two and three dimensions', *The Journal of the Acoustical Society of America*, vol. 138, no. 4, pp. 2326–2336, Oct. 2015. https://doi.org/10.1121/1.4931445

[2] M. Ryzy, T. Grabec, P. Sedlák, and I. A. Veres, 'Influence of grain morphology on ultrasonic wave attenuation in polycrystalline media with statistically equiaxed grains', *The Journal of the Acoustical Society of America*, vol. 143, no. 1, pp. 219–229, Jan. 2018. https://doi.org/10.1121/1.5020785

[3] M. Huang, G. Sha, P. Huthwaite, S. I. Rokhlin, and M. J. S. Lowe, 'Maximizing the accuracy of finite element simulation of elastic wave propagation in polycrystals', *The Journal of the Acoustical Society of America*, vol. 148, no. 4, pp. 1890–1910, Oct. 2020. https://doi.org/10.1121/10.0002102

[4] V. Dorval, N. Leymarie, A. Imperiale, E. Demaldent, Z. Aghenzour, and P.-E. Lhuillier, 'Determining ultrasonic propagation effective properties in complex heterogeneous media through microstructure-scale simulation', *Research and Review Journal of Nondestructive Testing*, vol. 1, no. 1, Aug. 2023. https://doi.org/10.58286/28132

[5] T. Yu, J.-F. Chaix, L. Audibert, D. Komatitsch, V. Garnier, and J.-M. Hénault, 'Simulations of ultrasonic wave propagation in concrete based on a two-dimensional numerical model validated analytically and experimentally', *Ultrasonics*, vol. 92, pp. 21–34, Feb. 2019. https://doi.org/10.1016/j.ultras.2018.07.018

[6] R. Hill, 'The Elastic Behaviour of a Crystalline Aggregate', *Proceedings of the Physical Society. Section A*, vol. 65, no. 5, pp. 349–354, May 1952. https://doi.org/10.1088/0370-1298/65/5/307

[7] W. Voigt, *Lehrbuch Der Kristallphysik*. Leipzig: Teubner, 1928.

[8] A. Reuss, 'Berechnung der Fließgrenze von Mischkristallen auf Grund der Plastizitätsbedingung für Einkristalle .', *ZAMM - Journal of Applied Mathematics and Mechanics / Zeitschrift für Angewandte Mathematik und Mechanik*, vol. 9, no. 1, pp. 49–58, Jan. 1929. https://doi.org/10.1002/zamm.19290090104

[9] H. M. Ledbetter and M. W. Austin, 'Texture in stainless steel welds: an ultrasonic study', *Journal of Materials Science*, vol. 20, no. 5, pp. 1720–1724, May 1985. https://doi.org/10.1007/BF00555276

Journal of Physics: Conference Series **2904** (2024) 012006 doi:10.1088/1742-6596/2904/1/012006

Impact of resynchronization errors on the localization quality of defects in reverberant plates

Omar Bouchakour, Lynda Chehami and Emmanuel Moulin

Université Polytechnique des Hauts-de-France, CNRS, UMR 8520 - IEMN - Institut d'Electronique de Microélectronique et de Nanotechnologie, F-59313 Valenciennes, France

E-mail: omar.bouchakour@uphf.fr

Abstract. In the context of Structural Health Monitoring (SHM), recent research works have been carried out to take advantage of the correlation of ambient noise to locate and characterize defects in reverberant plates. The back-propagation-based imaging algorithms developed in this way require that the signals recorded by the sensors of the control network be perfectly synchronized. This implies that these sensors must be linked to a common clock to ensure their synchronization, which generates significant complexity of the monitoring system in the case of embedded applications. Post-processing resynchronization is a promising solution to deal with this problem. However, in less favorable cases and especially in the presence of additive noise, this resynchronization could be made with some errors. In this work, we are interested in quantifying statistically the degradation of the contrast of localization images as a function of the statistical parameters of resynchronization errors. The obtained results are promising and can be of great interest for embedded applications.

1. Introduction

The need to monitor the structural state of infrastructures using systems with low energetic and hardware resources has given rise to so-called "passive" imaging techniques that take advantage of ambient noise to locate defects in reverberant mediums. These techniques, which require no controlled emissions, guarantee significant energy savings throughout the monitoring process. The works presented in[1, 2] have shown that it is possible to reconstruct the Green's function of a given medium using the cross-correlations between a set of sensors recording the ambient noise to which the medium is subjected.

Recently, the field of Structural Health Monitoring (SHM) has undergone a significant evolution with the growing adoption of autonomous sensor networks integrated directly into the structures themselves [3, 4]. This innovation aims to transform the way we monitor the integrity of infrastructures by enabling continuous and real-time monitoring. The use of these embedded sensors reduces the complexity of traditional monitoring systems and increases their efficiency, as they provide accurate and immediate data on the condition of structures. This enables engineers and maintenance managers to make rapid and informed decisions, minimising the risk of structural failure and extending the life of installations.

Although independent sensor networks dedicated to SHM have demonstrated a significant reduction in cost and complexity, resynchronization remains a major challenge for this technology [5, 6]. In fact, since each receiver operates with its own independent internal clock, significant drifts occur between the recorded signals. To remedy this, a lot of synchronization methods have been proposed in the literature [7, 8, 9]. However, it should be noted that the most precise ones require the use of additional hardware or communication protocols so that the sensor network is well synchronized. Nevertheless, this contradicts the idea of developing autonomous monitoring networks with low material and energy resources.

The Peak Correlation Technique (PCT) [10, 11, 12] is recognized as promising method for detecting time shifts between signals. Basically, it involves correlating the positive and negative time lags of the cross-correlation between signals recorded by a pair of sensors. A peak in this correlation corresponds to the time shift between these signals, facilitating sensor resynchronization in post-processing without the need for additional electronic hardware complexity. However, the presence of noise in the measurement process can lead to imperfect resynchronization and subsequently poor defect localization.

This work is outlined as follows : we initially focus on theoretically quantifying the impact of resynchronization errors on the quality of defect detection in elastic reverberating plates. Subsequently, we conduct some numerical simulations to validate the theoretical findings.

2. Theoretical developments

We consider a configuration of N independent acoustic sensors placed on a plate exposed to some ambient noise (see Fig. 1). The methodology for detecting and localizing potential defects involves cross-correlating the noise recorded by each sensor, coupled with the implementation of a back-propagation and dispersion-compensation algorithm developed in [13, 14]. Indeed, by computing the positive time lag in differential noise cross-correlation between two cases–with and without a defect–it is possible to isolate the defect signature. Subsequently, a pixel-by-pixel back-propagation technique is employed to construct an image that accurately localizes the defect.

We assume that each receiver operates with its own independent internal clock, which can result in potential time drifts between recorded signals. Consequently, resyn-

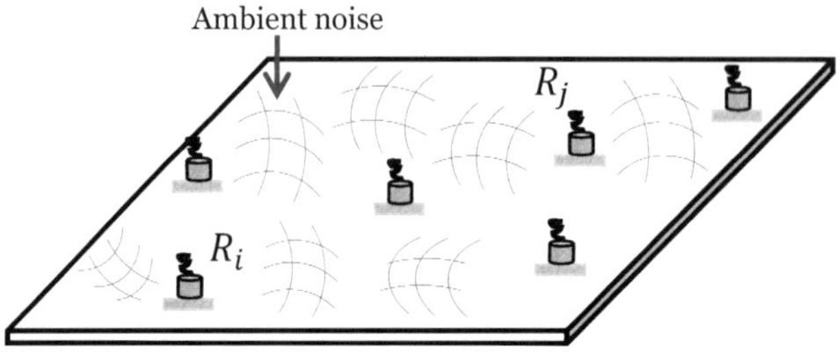

Figure 1: Schematic representation of the studied configuration.

chronization of the signals is essential before applying the previously explained localization algorithm.

Here we are interested in quantifying the degradation of the defect detection as a function of resynchronization errors. For this, we define a contrast ratio denoted C_r between the case with resynchronization error and a reference case without resynchronization error as

$$\mathcal{C}_r = \frac{\langle I_{res} \rangle}{I}, \tag{1}$$

where I_{res} and I are the defect pixel intensities in the cases of imperfectly and perfectly resynchronized signals respectively and $\langle . \rangle$ denotes the mathematical expectation on a set of realizations on synchronization errors.

In a previous work [15], we have shown that in the case of small values of the standard deviation of resynchronization errors, the contrast ratio can be written as

$$\mathcal{C}_r \approx 2\pi |p(\omega_0)|^2, \tag{2}$$

where $p(\omega_0)$ is the probability density function of resynchronization errors at ω_0 the central frequency of the bandwidth.

This formulation demonstrates the relationship between localization performance and the standard deviation of the resynchronization errors. This relationship will be applied and validated through a numerical example in the next section.

3. Numerical validation

In order to validate the previously obtained theoretical results, a numerical simulation is carried out in this section using a finite element code ELMER. This code simulates acoustic wave propagation in a finite plate using the Reissner-Mindlin plate model.

The configuration tested (see Fig 2) consists of an aluminum plate of dimensions 1×0.5 m^2 and thickness 3 mm, its physical properties are noted in table 1.

Journal of Physics: Conference Series **2904** (2024) 012006 doi:10.1088/1742-6596/2904/1/012006

Table 1: Aluminum's physical properties: density ρ, Young's modulus E and Poisson's ratio ν.

Material	ρ $(kg.m^{-3})$	E (GPa)	ν
Aluminum	2700	70	0,35

On the surface of this plate, we place an array of 8 sensors at known positions and a distribution of 600 sources uniformly distributed around the sensors. A defect is modeled by a point contact normal to the surface of the plate at position (53, 26) cm.

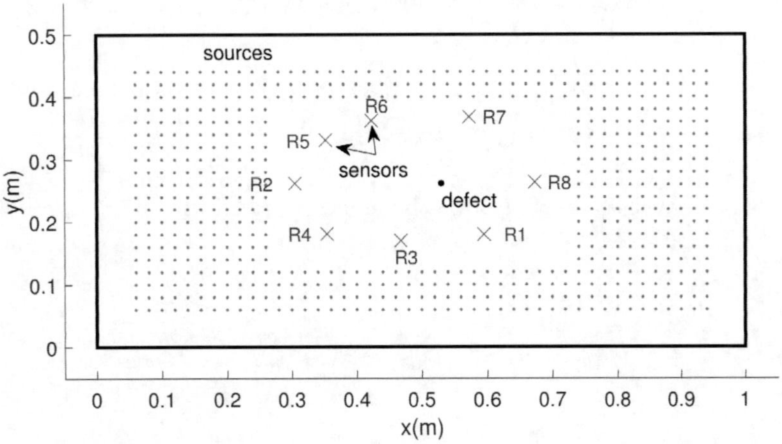

Figure 2: Simulated configuration: aluminum plate 1×0.5 m^2 and 3 mm thick. 8 sensors (blue crosses). 600 sources uniformly distributed around the sensors (red dots). A point defect at position (53, 26) cm.

The emission signal is a Gaussian-windowed sine of frequency $f_0 = 20$ kHz. The sampling frequency is 2.5 MS/s. For each pair of receivers, the summed cross-correlation over all sources is computed.

The back-propagation-based imaging algorithm previously recalled is then applied on the positive time lag of computed differential cross-correlations. The obtained localization image is shown in Fig. 3. Pixel intensities are normalized with respect to the maximum intensity. As expected, the defect location is correctly localized.

Now we are interested in highlighting the effect of synchronization errors on the defect localization image quality. These errors are modeled by random shifts following a Gaussian distribution $\mathcal{N}(0, \sigma)$ introduced on the computed cross-correlations. For each value of σ, the contrast ratio between the cases with and without synchronization errors

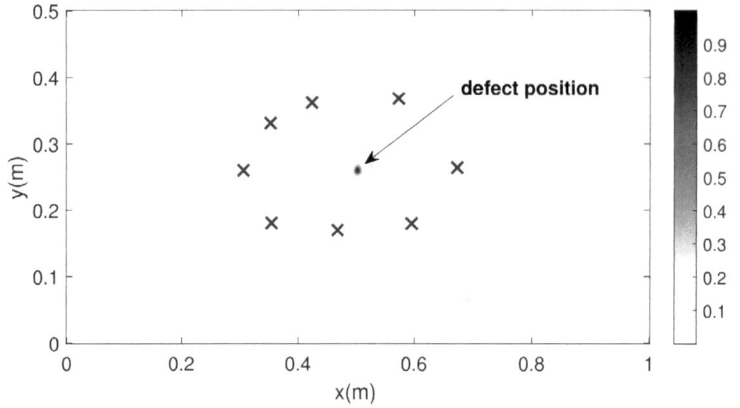

Figure 3: Defect localization image.

is computed and averaged over 500 realizations. This averaged numerical contrast ratio can then be compared to the theoretical expression established in Sec. 2.

In the case of a Gaussian distribution, the Fourier transform of the probability density function is

$$p(\omega) = \frac{1}{\sqrt{2\pi}} e^{\frac{-\sigma^2\omega^2}{2}}. \tag{3}$$

Consequently, Eq. 2 in Sec. 2 becomes

$$\mathcal{C}_r = e^{-\sigma^2\omega_0^2}. \tag{4}$$

The corresponding curves as a function of σ are represented in Fig. 4, along with the averaged numerical contrast ratio computed from the simulation.

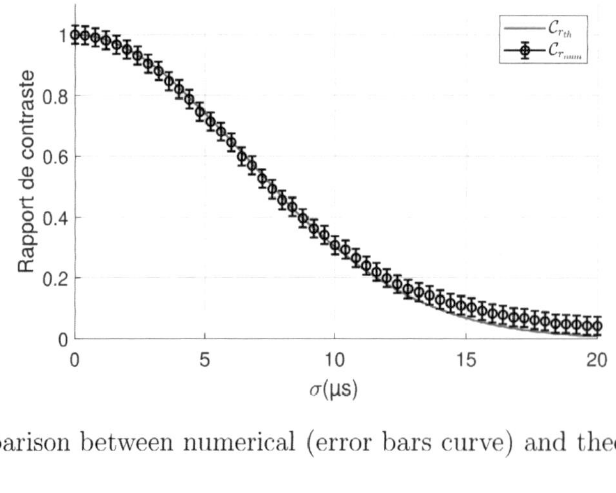

Figure 4: Comparison between numerical (error bars curve) and theoretical (solid line) contrast ratios.

Journal of Physics: Conference Series **2904** (2024) 012006 doi:10.1088/1742-6596/2904/1/012006

Referring to Fig. 4, the numerical contrast ratio matches the theoretical prediction for small values of σ, ($\sigma \leq 12\ \mu s$). Above 12 μs, we notice a discrepancy of the numerical curve compared to the theoretical one. This is due to the assumption of low sigma values considered in the theoretical developments. The error bars are estimated from the fluctuations of the contrast ratio with respect to the realizations on the synchronization errors. The small error bars indicate that the contrast ratio does not fluctuate significantly across realizations and can be favorably compared with the developed theoretical formula [Eq. 4].

4. Conclusion

This paper addresses the issue of potentially imperfect signal resynchronization in passive imaging using the PCT in plate-like structures, particularly when employing acoustic sensors with independent time-bases. First, we have developed a theoretical formula to quantify how the contrast ratio in defect localization images deteriorates as a function of the statistical properties of resynchronization errors in correlation functions. Then, a finite element simulation was conducted in order to validate the relationship between the contrast ratio and the standard deviation of resynchronization errors.

5. References

[1] O. I. Lobkis and R. L. Weaver. On the emergence of the Green's function in the correlations of a diffuse field. *J. Acoust. Soc. Am.*, **110**:3011–3017, 2001.

[2] A. Colombi, L. Boschi, P. Roux, and M. Campillo. Green's function retrieval through cross-correlations in a two-dimensional complex reverberating medium. *J. Acoust. Soc. Am.*, **135**(3):1034–1043, 2014.

[3] T. Buckley, B. Ghosh, and V. Pakrashi. Edge structural health monitoring (e-shm) using low-power wireless sensing. *Sensors*, 21(20):6760, 2021.

[4] A. Sofi, J. Regita, B. Rane, and H. H. Lau. Structural health monitoring using wireless smart sensor network–an overview. *Mech. Syst. Signal Process.*, 163:108113, 2022.

[5] A. Araujo, J. García-Palacios, J. Blesa, F. Tirado, E. Romero, A. Samartín, and O. Nieto-Taladriz. Wireless measurement system for structural health monitoring with high time-synchronization accuracy. *IEEE Trans. Instrum. Meas.*, **61**(3):801–810, 2011.

[6] C. B. Nzouatchoua, M. Bentahar, S. Montresor, N. Colin, V. Le Cam, C. Trottier, and N. Terrien. Damage localization on composite structures based on the delay-and-sum algorithm using simulation and experimental methods. *Sensors*, **23**(9):4368, 2023.

[7] O. Montenbruck, M. Garcia-Fernandez, and J. Williams. Performance comparison of semicodeless gps receivers for leo satellites. *GPS Solut.*, **10**:249–261, 2006.

[8] J. Elson, L. Girod, and D. Estrin. Fine-grained network time synchronization using reference broadcasts. *ACM SIGOPS Oper. Syst. Rev.*, **36**(SI):147–163, 2002.

[9] M. Maróti, B. Kusy, G. Simon, and A. Lédeczi. The flooding time synchronization protocol. In *Proc. 2nd Int. Conf. Embedded Netw. Sens. Syst.*, pages 39–49, 2004.

[10] K. G. Sabra, P. Roux, A. M. Thode, G. L. D'Spain, WS. Hodgkiss, and WA. Kuperman. Using ocean ambient noise for array self-localization and self-synchronization. *IEEE Journal of Oceanic Engineering*, **30**(2):338–347, 2005.

[11] C. Sens-Schönfelder. Synchronizing seismic networks with ambient noise. *Geophys. J. Int.*, **174**(3):966–970, 2008.

[12] L. Tenorio-Hallé, A. M. Thode, S. L. Swartz, and J. Urbán R. Using anisotropic and narrowband ambient noise for continuous measurements of relative clock drift between independent acoustic recorders. *J. Acoust. Soc. Am.*, **149**(6):4094–4105, 2021.

[13] L. Chehami, E. Moulin, J. de Rosny, C. Prada, O. Bou Matar, F. Benmeddour, and J. Assaad. Detection and localization of a defect in a reverberant plate using acoustic field correlation. *J. Appl. Phys.*, **115**(10):104901, 2014.

[14] L. Chehami, J. de Rosny, C. Prada, E. Moulin, and J. Assaad. Experimental study of passive defect localization in plates using ambient noise. *IEEE Trans. Ultrason., Ferroelectr., Freq. Control*, **62**(8):1544–1553, 2015.

[15] O. Bouchakour, E. Moulin, L. Chehami, and N. Smagin. Quantification and mitigation of the effect of resynchronization errors in ultrasound sensor network for passive imaging in elastic plates. *J. Acoust. Soc. Am.*, **155**(5):3283–3290, 2024.

Determination of ageing indicators on glass-fiber polyester composite skins using Lamb guided waves

Khalid AOUJDAD(1), Mounsif ECH-CHERIF EL-KETTANI(1), Damien LEDUC(1), Pierre MARECHAL(1),

(1) Laboratory of Waves and Complex Media (LOMC UMR 6294), Le Havre Normandy University, Le Havre, France

Khalid.aoujdad@doct.univ-lehavre.fr

Abstract

This paper investigates the structural integrity of glass fibre reinforced polymer (GFRP) composites used in offshore wind turbine blades.

Samples of two thicknesses (4-plies and 6-plies) undergo accelerated ageing inside tanks full of seawater of salinity 28-30% heated at 40°C, whereas non-aged samples remain as a reference.

Ultrasonic non-destructive evaluation (UNDE) using Lamb guided waves is applied to assess the behaviour of samples in those conditions based on ageing indicators investigation. The findings indicate a trend toward an increase in wavenumber and a decrease in phase velocity of Lamb waves modes for aged samples, which suggests a decrease in mechanical properties over time. Furthermore, this disparity becomes noticeably more evident with longer ageing times.

Keywords

Offshore wind turbine blades (OWTB), ultrasonic non-destructive evaluation (UNDE), Lamb guided waves (LGW), glass fibre reinforced polymer (GFRP), composite material, accelerated ageing.

1. Introduction

Offshore wind power is a highly promising path to sustainable electricity production. However, the environmental operating conditions, i.e. high levels of moisture, salt, and temperature fluctuations of marine environment, have a significant impact on the structural integrity and performance of blade materials over time. Therefore, ensuring the long-term durability and reliability of offshore wind turbine blades (OWTB) remains a major challenge. Therefore, the present research work focuses on the complex relationships between material structure, properties and acoustic responses of OWTB materials. By employing ultrasonic non-destructive evaluation (UNDE) methods, our aim is to gain comprehensive insights into the degradation mechanisms and durability of these vital components.

The performance, durability and reliability of the OWTB [1] are highly dependent on the mechanical properties of their material. Glass fibre reinforced polymers (GFRP) are widely used due to their exceptional weight/performance ratio and fatigue resistance [2]. Nevertheless, the use of composite materials in OWTB requires the assessment of environmental resistance to factors such as humidity and temperature [3]. In addition, assessment of manufacturing processes and costs helps optimize blade performance and profitability [4]. By simulating ageing in sea medium, we aim to replicate the damaging effects of prolonged exposure to moisture and salt on blade materials. Previous studies have consistently shown that GFRP are basically more resistant to seawater corrosion compared to other types of fibre-reinforced composites [5]. Studies on the hygrothermal effects should primarily involve the influence of moisture and temperature. Regarding moisture, mechanical assessments of the impact of water absorption on residual mechanical properties reveal that each type of composite exhibits a distinct water concentration threshold, beyond which a significant reduction in mechanical properties occurs [6]. As far as temperature assessment is concerned, the glass transition temperature Tg is essential for understanding both accelerated ageing studies and to ensure the long-term health and durability of materials [7].

The effects of accelerated seawater ageing are evaluated using ultrasonic guided waves. Numerous non-destructive testing (NDT) methods have been widely used and evaluated for the structural health monitoring (SHM) of wind turbine blades. Civera and Surace [8] have reviewed over 300 documents published in the last two decades and concluded that a combination of both contact and non-contact is the most effective approach for SHM. The guided waves (GW) methods in structural health monitoring (SHM) show promising results in detecting significant flaws in turbine components. Gan et al. [9] highlighted the strategic importance of monitoring turbine blades for preventative maintenance. However, challenges remain, particularly in the use of GW techniques, due to material anisotropy and manufacturing process. T. Agra [10] carried out her PhD work on guided Lamb waves, showing their robust sensitivity to structural changes caused by ageing. By analysing the propagation velocities of GW through samples, subtle changes in material properties can be detected, such as density, longitudinal and transverse stiffness moduli. These acoustic indicators provide valuable clues for assessing the extent of material degradation, leading to the identification of early signs of structural weakness or damage.

In this paper, the characterisation of OWTB is presented by means of Lamb waves to evaluate the effect of ageing in marine environment. In section 2, the reference sample properties, accelerated ageing and water uptake intake characteristics are presented. The third and fourth sections are dedicated to UNDE and wavenumber dispersion curves analysis, respectively.

2. UD-GFRP composite samples

After carrying out a comparative analysis of the materials used in blade fabrication, a visit to LM Wind Power in Cherbourg enabled to fix the sample manufacturing process. The studied sandwich composite samples are composed of two skins that are made of unidirectional glass fibre reinforced polyester (UD-GFRP), and a styrene-acrylo-nitrile (SAN) foam core. The ultrasonic characterization focuses on the skin part including both non-aged and aged samples of 4 and 6 plies. The two samples are heterogeneous and may behave differently due to the non-proportional thickness between the 4-plies (2 mm), considered fibre-dominant composite, and the 6-plies (4 mm), considered resin-dominant composite. This leads to differences in densities and elastic constants.

The mechanical properties of the composite samples are summarised in Table 1.

Thickness [mm]	*Density* [kg.m^{-3}]	*CL* [m.s^{-1}]	*CT* [m.s^{-1}]
2 (4 plies) – **4** (6 plies)	1700	2730	1430

Table 1. Mechanical properties of un-aged UD-GFRP samples.

2.1. Accelerated ageing

The marine environment is mimicked by hygrothermal ageing of GFRP samples. A dedicated setup is built to immerse the samples in seawater heated to 40°C for periods of 15 and 28 weeks. This experimental methodology provides a method for studying the progressive deterioration of material characteristics over time and its implications for structural integrity.

Figure 1. Pictures of the non-aged (left) and aged (right) UD-GFRP composite skin.

As illustrated by Figure 1, the opacity of the aged sample compared to non-aged increases after removal from the immersion tank dedicated to ageing. Which can be explained by the chemical degradation of the polyester resin matrix and the appearance of micro-cracks and small

Journal of Physics: Conference Series **2904** (2024) 012007 doi:10.1088/1742-6596/2904/1/012007

crystalline zones (crystallites), which reduces the transparency of the sample. This degradation is due to a modification of the polymer chains and plasticisation of the polyester resin matrix.

2.2. Moisture behaviour

As illustrated in Figure 2, the moisture intake of the samples is studied by measuring the mass evolution over the immersion time.

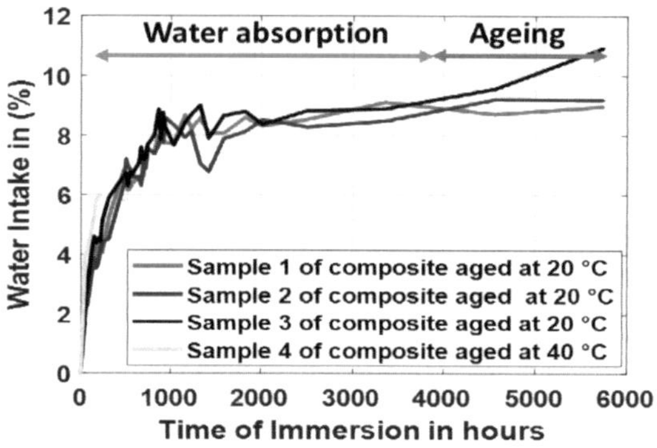

Figure 2. Water intake curves of the samples 1 to 4

The curve exhibits a rapid growth, reaching a maximum value and stabilizing thereafter. Over a period of 12 weeks (~2000 h), the materials reach water saturation. During this period, materials gradually absorb water from their environment, up to 9% of their total weight. Once saturated, the ageing process begins, and seawater begins to attack the material structure.

3. Ultrasonic Non-Destructive Characterisation

3.1. Experimental setup

In the experimental setup, ultrasonic guided waves (called Lamb or plate waves) are used as an efficient tool for the comprehensive evaluation of the properties and behaviours of composite materials. These guided waves are generated using a contact transducer with a central frequency of 1 MHz, angled at 30° using a Plexiglas wedge (Figure 3).

The transducer is excited by an *Olympus* 5058PR pulse generator. A *Polytec* laser velocimeter (*VibroFlex Qtec* infrared) is used as a receiver, detecting displacements normal to the free surface of the sample. These displacements of the wave as a function of time are recorded in the main direction of propagation (along the fibres direction) over a distance of 50 mm in steps of 0.1 mm, i.e. 500 positions. This step is adapted to the wavelength of the waves generated in the material, which is around 2.4 mm. To follow the propagation of the Lamb mode generated, the sample is shifted vertically (keeping the laser velocimeter stationary). Vertical movement along the Z axis is provided by a motorised support controlled by a *Newport XPS* motion controller, to which the sample is attached. Due to the transparent structure of the sample, a reflective film is added to its surface to ensure maximum reflection towards the laser source (Figure 4).

Figure 3. Picture of the experimental setup of Laser.

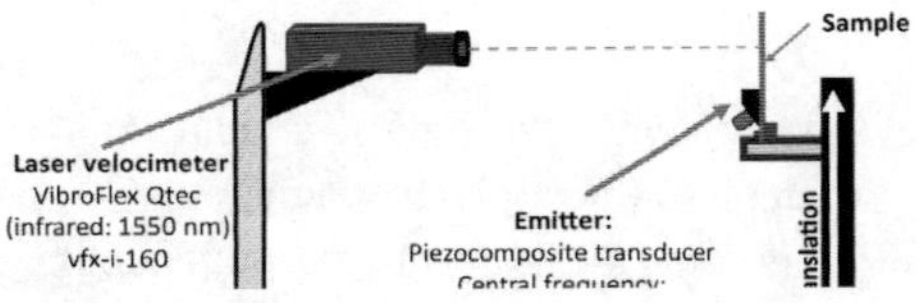

Figure 4. Schematic representation of the laser experimental setup used for capturing ultrasonic guided waves in the composite material

3.2. Signal acquisition and processing

The experimental phase consisted of collecting raw data of Lamb waves from non-aged and aged samples over a time range of 200 µs. To ensure accuracy and minimise noise or unwanted signals, the data is pre-processed using by averaging (128 averages). Next, a double Fourier transform is applied, converting the data from the time-position domain to the frequency-wavenumber domain (Figure 5). An effective wavenumber resolution of $k_{step} = 3{,}8\ m^{-1}$ is achieved through zero-padding, while the actual wavenumber resolution without zero-padding is approximately $125.6\ m^{-1}$. This process reveals the wavenumber as a function of the frequency components of Lamb waves, where the amplitudes of the modes are represented using the colour scale.

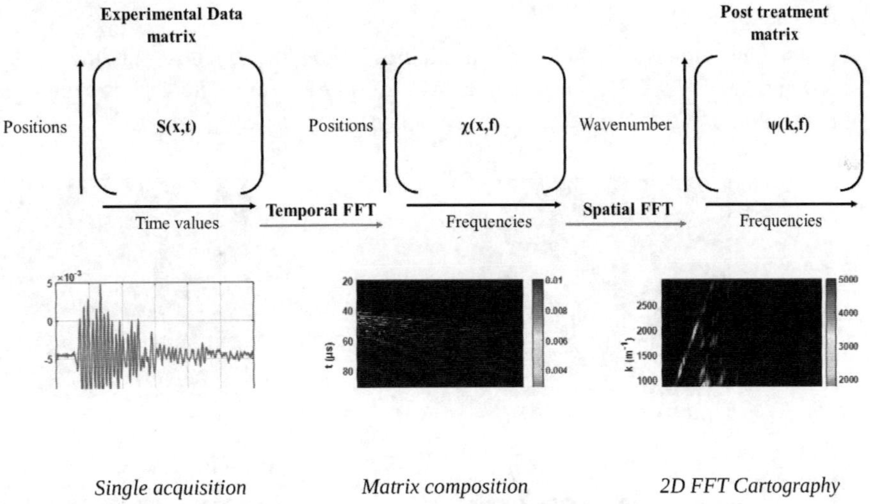

Figure 5. Signal processing workflow for ultrasonic Lamb wave data.

4. Wavenumber and bandwidth variations as ageing indicator

Variations in wavenumber before and after ageing provide an insight into changes in wave propagation properties, reflecting alterations in the internal structure and indicating degradation. In this study of the samples aged at 40 °C, we assume that the effects on material thickness are not considered. This quantitative assessment facilitates preventive maintenance and ensures the durability of composite structures. The second study will explore the impact of ageing on frequency response, with changes in viscosity affecting signal bandwidth, providing a means of detecting ageing-induced alterations in composite materials.

Journal of Physics: Conference Series **2904** (2024) 012007 doi:10.1088/1742-6596/2904/1/012007

4.1. About the wavenumber

The wavenumber k is of crucial interest for assessing the effects of ageing on composite materials. For a given frequency, the phase velocity c can be extracted since $c = \omega/k$. By monitoring variations in the wavenumber positions (Figure 6) before and after ageing, we can detect changes in the wave propagation properties induced by ageing. These changes can reveal alterations in the internal structure of the material, such as the appearance of micro-cracks, delamination or variations in stiffness, all of which are signs of degradation to which guided waves are sensitive [11]. Consequently, wavenumber analysis provides a quantitative assessment of the extent and nature of damage caused by ageing, offering valuable information for preventive maintenance and the durability of composite structures.

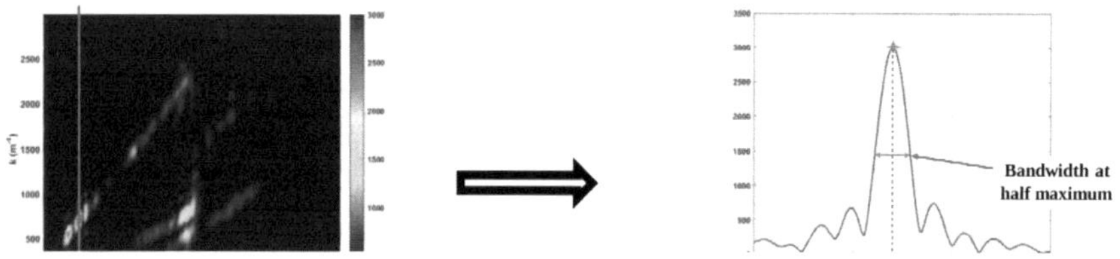

Figure 6. Band extraction from a frequency slice of the 2D FFT, as a function of wavenumber
The colour scale represents the amplitude of the Lamb modes.

4.2. Comparative analysis

Four studied samples are illustrated in Figure 7, showing the corresponding 2D FFT of non-aged and aged UD GFRP, 4 plies above and 6 plies below. The 4 plies samples exhibit the propagation of 3 Lamb modes, while the 6 plies samples show 4 modes.

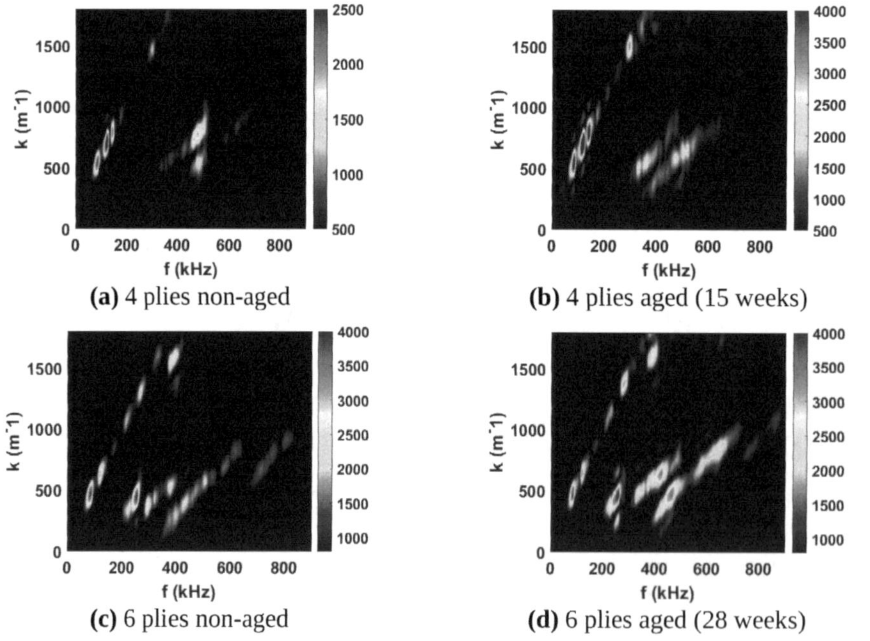

Figure 7. 2D FFT of 4 plies samples above, and 6 plies samples below

4.2.1. Wavenumber variations

An analysis of the wavenumber evolution is used as an investigation tool. Since $c = 2.\pi.f/k$, the wavenumber evolution is directly linked to the change in phase velocity. A comparison between the wavenumbers of non-aged (*kNA*) and aged (*kA*) samples is presented in the tables shown in Figure 8 and Figure 9, for 4 plies and 6 plies samples, respectively. Only well-defined wave spectra can be used in this study, which allows us to consider a well-generated Lamb mode in the non-aged sample and its equivalent in the aged sample.

As illustrated, the wavenumber evolutions exhibit generally rightward shifts $\Delta k > 0$, indicating a reduction of phase velocity due to ageing. However, it is normal that some peaks (typically not the peak associated to the mode of interest) show a negligible shift to the left, likely due to spectral leakage caused by signal windowing. These peaks may be caused by noise or disturbances in the original signal. Finally, if these shifts are important, they may be attributed to the polyester matrix damages that increases the stiffness of the sample making it less flexible and more brittle after ageing.

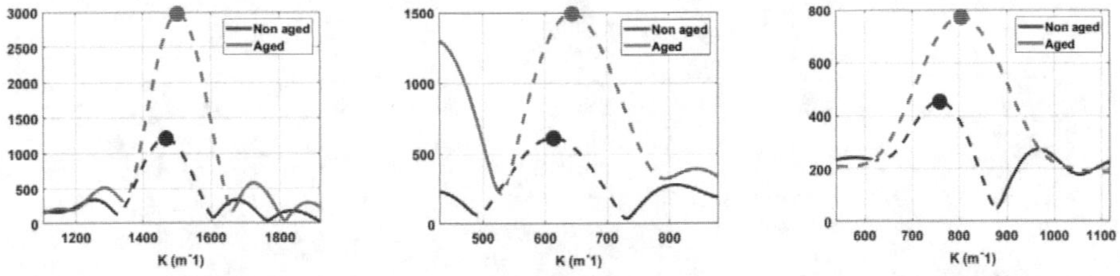

f = 300 kHz

$K_{NA}(m^{-1})$	$K_A(m^{-1})$	$\Delta k(m^{-1})$
1465	1503	38

f = 400 kHz

$K_{NA}(m^{-1})$	$K_A(m^{-1})$	$\Delta k(m^{-1})$
613.6	644.3	30.7

f = 600 kHz

$K_{NA}(m^{-1})$	$K_A(m^{-1})$	$\Delta k(m^{-1})$
759.3	805.3	46

Figure 8. Wavenumber shifts of 4 plies samples

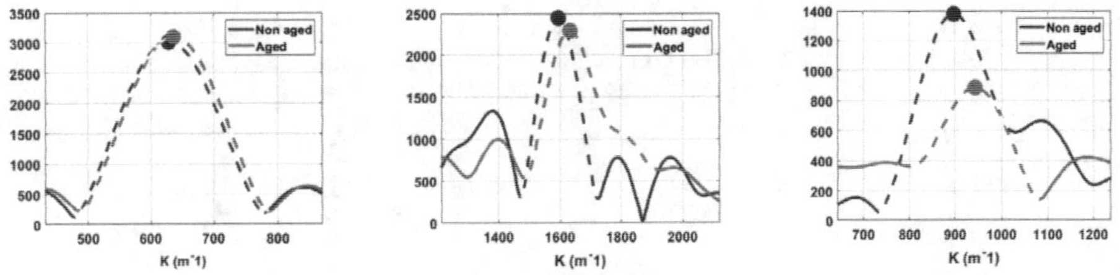

f = 120 kHz

$K_{NA}(m^{-1})$	$K_A(m^{-1})$	$\Delta k(m^{-1})$
628.9	636.6	7.7

f = 400 kHz

$K_{NA}(m^{-1})$	$K_A(m^{-1})$	$\Delta k(m^{-1})$
1595	1634	38.4

f = 800 kHz

$K_{NA}(m^{-1})$	$K_A(m^{-1})$	$\Delta k(m^{-1})$
897.4	943.4	46

Figure 9. Wavenumber shifts of 6 plies samples

The comparison between the behaviour of 4 plies samples (Figure 8), and 6 plies samples (Figure 9) shows a similar overall response. This concordance confirms the reliability of wavenumber as an indicator of ageing, since it accurately reflects changes in phase velocity that are sensitive to internal degradation. This conclusion assumes that the thickness of the samples remains constant throughout the ageing process, and the changes phase velocity are due to mechanical properties degradation.

Journal of Physics: Conference Series **2904** (2024) 012007 doi:10.1088/1742-6596/2904/1/012007

4.2.1.1. Variation rate of the wavenumber

The comparison of wavenumber variation rates $\Delta k/k_{NA}$ between non-aged and aged samples was done on the same mode at different frequency. The differences are calculated on the relevant peaks at each frequency in both the non-aged and aged samples with an estimated uncertainty of $\pm 2\%$.

The variation rates are shown in the following charts:

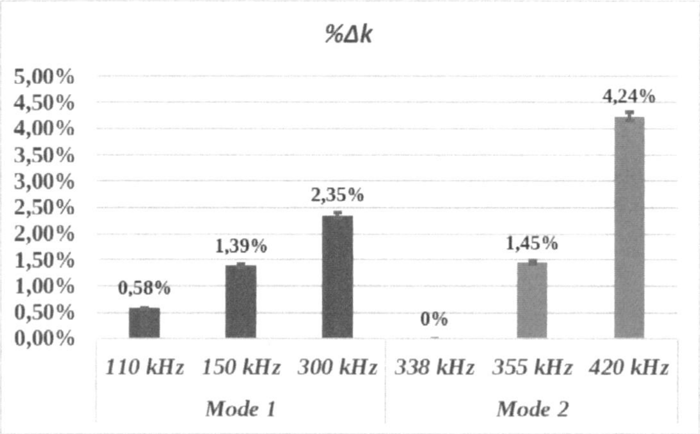

Figure 10. Variation rates of the wavenumber on GFRP 4 plies

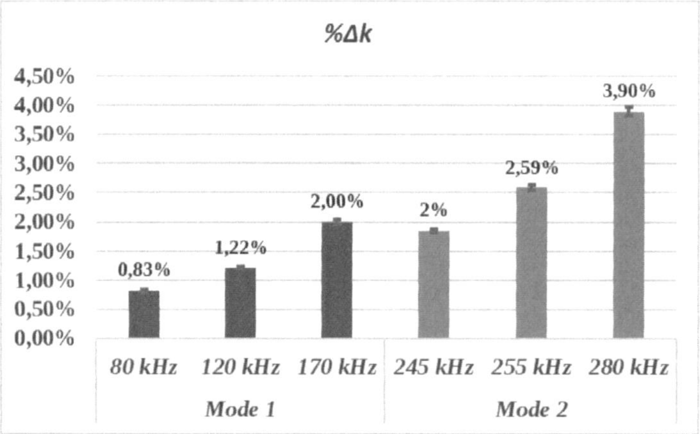

Figure 11. Variation rates of the wavenumber on GFRP 6 plies

A notable disparity in wavenumber variation rates is observed between the 4 plies samples (Figure 10) and the 6 plies samples (Figure 11). For both modes, the variations in wavenumber increase with frequency because Lamb waves at high frequencies are more sensitive to material degradation, such as reductions in elastic parameters, which increase these shifts. In addition, the increase in material attenuation with frequency may contribute to this amplification. The variations rates of the wavenumber between the 4 plies samples and 6 plies vary according to mode (1 or 2). These results suggest that the ageing process interacts differently with the material structure in each configuration, leading to distinct responses to environmental solicitations.

Journal of Physics: Conference Series **2904** (2024) 012007 doi:10.1088/1742-6596/2904/1/012007

4.2.2. Bandwidth variations

In this second study, the effect of ageing is studied on the bandwidth of the signal. Research has shown that changes in the viscosity of the medium in which guided waves propagate can influence the bandwidth of the signal. A higher viscosity tends to damp the wave and broaden its content, resulting in an increase in bandwidth [12]. Therefore, the differences in bandwidth at half maximum (BWHM) between non-aged and aged samples may indicate changes in viscosity resulting from the ageing process of the composite materials. In summary, the greater the bandwidth of a wave, the more attenuating the medium tends to be, as it indicates a greater range of frequencies or wavelengths contributing to energy dissipation through mechanisms such as absorption and scattering [13].

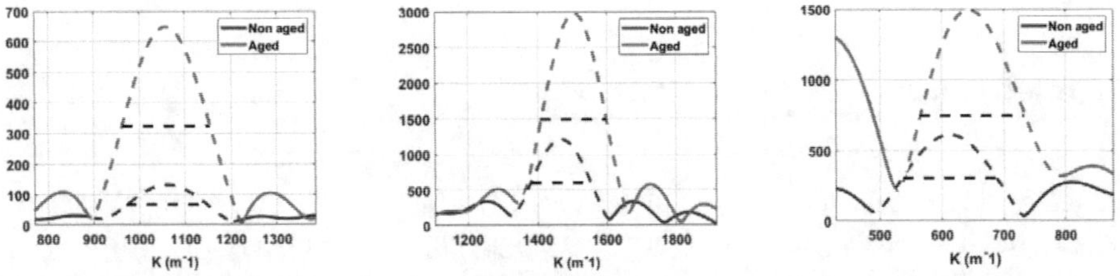

f = 200 kHz f = 300 kHz f = 400 kHz

Figure 12. Bandwidths of 4 plies samples.

Frequency [kHz]	*200*	*300*	*400*
BWHM (non-aged) [kHz]	176.8	172	153.4
BWHM (aged) [kHz]	191.4	192	168.8
Δ_{BWHM} [kHz]	14.6 (8%)	20 (12%)	15.4 (10%)

Table 2. Bandwidth gaps and variation rates of 4 plies samples

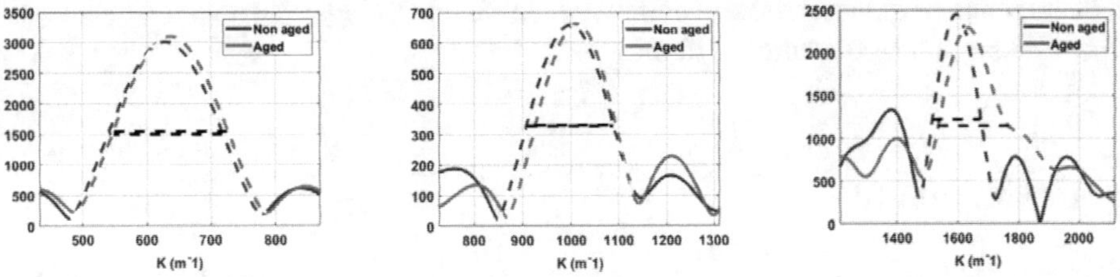

f = 120 kHz f = 200 kHz f = 400 kHz

Figure 13. Bandwidths of 6 plies samples.

Frequency [kHz]	*120*	*200*	*400*
BWHM (non-aged) [kHz]	177	180	161
BWHM (aged) [kHz]	177	165	224
Δ_{BWHM} [kHz]	0 (0%)	-15 (-8.3%)	63 (+39%)

Table 3. Bandwidth gaps and variation rates of 6 plies samples

As illustrated by the evolutions of the resonances bandwidths, significant variation rates can be observed both for 4 plies samples (Figure 12 and Table 2) and 6 plies samples (Figure 13 and Table 3). Nevertheless, the second mode observed at f = 200 kHz do not present a similar trend, since the indicator Δ_{BWHM} decreases in this specific case. The detailed study of this particular mode needs further investigation in future works. The measurement results indicate that the ageing may be clearly discriminated in the two studied configurations, either 4 plies or 6 plies samples.

5. Conclusion

In summary, an experimental study explores the use of Lamb guided waves as a characterize the ageing of UD-GFRP composites. The results highlight the sensitivity of wavenumbers and signal bandwidths to the effects of the ageing process, demonstrating that these parameters provide robust and reliable indicators of material degradation. More specifically, the rate of changes observed in wavenumbers and bandwidths suggest the potential of a quantitative tool to assess the effects of ageing. However, further studies and analyses are required to establish a definitive link between these changes and material properties.

Within 30 weeks at 40°C, the ageing effects observed are limited. This indicates that material performance is relatively stable over these conditions. However, ongoing studies involving longer immersion periods and higher temperatures (typically 60°C) are in progress to assess long-term durability and potential internal degradation of the samples.

Further research on additional ageing indicators could provide valuable insights into the materials ageing dynamics of these materials, and inform on strategies to improve their stability and performance under offshore conditions.

Acknowledgments

The authors acknowledge the financial support of the French Agence Nationale de la recherche LabEx EMC3 through the project ... (Grant No. 10-LABX-0009), and the Normandy Region (Réseau d'Intérêt Normand - Label d'excellence).

References

[1] A. Chehouri, R. Younes, A. Ilinca, et J. Perron, « Review of performance optimization techniques applied to wind turbines », *Applied Energy*, vol. 142, p. 361-388, 2015, doi: 10.1016/j.apenergy.2014.12.043.

[2] L. Mishnaevsky, K. Branner, H. N. Petersen, J. Beauson, M. McGugan, et B. F. Sørensen, « Materials for Wind Turbine Blades: An Overview », *Materials*, vol. 10, n° 11, Art. n° 11, 2017, doi: 10.3390/ma10111285.

[3] K. Liao, C. R. Schultheisz, et D. L. Hunston, « Effects of environmental aging on the properties of pultruded GFRP », *Composites Part B: Engineering*, vol. 30, n° 5, p. 485-493, 1999, doi: 10.1016/S1359-8368(99)00013-X.

[4] U. E. Y. Gómez, Z. J. A. López, R. A. Jimenez, G. V. López, et L. J. J. Villalon, « Design and Manufacturing of Wind Turbine Blades of Low Capacity Using CAD/CAM Techniques and Composite Materials », *Energy Procedia*, vol. 57, p. 682-690, 2014, doi: 10.1016/j.egypro.2014.10.223.

[5] S. Wen et M. Cao, « Review on degradation behavior of fiber-reinforced polymer bars in marine environments », *Journal of Materials Science*, vol. 58, p. 1-33, 2023, doi: 10.1007/s10853-023-08667-1.

[6] D. Valentin, F. Paray, et B. Guetta, « The hygrothermal behaviour of glass fibre reinforced Pa66 composites: A study of the effect of water absorption on their mechanical properties », *J Mater Sci*, vol. 22, n° 1, p. 46-56, 1987, doi: 10.1007/BF01160550.

[7] J. Earl et R. Shenoi, « Hygrothermal ageing effects on FRP laminate and structural foam materials », *Composites Part A: Applied Science and Manufacturing*, vol. 35, n° 11, p. 1237-1247, 2004, doi: 10.1016/S1359-835X(04)00121-6.

[8] M. Civera et C. Surace, « Non-Destructive Techniques for the Condition and Structural Health Monitoring of Wind Turbines: A Literature Review of the Last 20 Years », *Sensors*, vol. 22, n° 4, Art. n° 4, 2022, doi: 10.3390/s22041627.

[9] K. Burnham, V. Dimlaye, S. Soua, et T.-H. Gan, « Real-time monitoring system for defects detection in wind turbine structures and rotating components. ».

[10] T. Agra, « Evaluation ultrasonore du vieillissement climatique de collages structuraux composite-composite ».

[11] S. Baste, R. El Guerjouma, et B. Audoin, « Effect of microcracking on the macroscopic behaviour of ceramic matrix composites: Ultrasonic evaluation of anisotropic damage », *Mechanics of Materials*, vol. 14, n° 1, p. 15-31, 1992, doi: 10.1016/0167-6636(92)90015-6.

[12] J. L. Rose, *Ultrasonic Guided Waves in Solid Media*. Cambridge: Cambridge University Press, 2014. doi: 10.1017/CBO9781107273610.

[13] L. Attar, M. Ech Cherif El Kettani, D. Leduc, M. V. Predoi, et J. Galy, « Detection of kissing bond type defects and evaluation of the bonding quality in metal/adhesive/composite structures by a wavenumber-frequency insensitive SH mode », *NDT & E International*, vol. 137, p. 102841, 2023, doi: 10.1016/j.ndteint.2023.102841.

Acoustic radiation force generated by vortex beams on microbubbles embedded in a soft hydrogel

Antoine Penneron, Thomas Brunet, Diego Baresch
Université de Bordeaux, CNRS, Bordeaux INP, ENSAM, INRAE, UMR 5295 I2M, F-33405 Talence, France
E-mail: antoine.penneron@u-bordeaux.fr, diego.baresch@u-bordeaux.fr

Abstract. Acoustical tweezers are recently developed experimental tools that enable the contactless manipulation of a wide range of particles in fluid media. Their good penetrability, high trapping strength and biocompatibility have poised these acoustic methods as interesting alternatives to other well established optical and magnetic manipulation techniques. This work describes preliminary steps in the development of an active approach to measure the local elasticity of soft hydrogels using single-beam acoustical tweezers. The displacement of a microbubble embedded in a hydrogel is induced by the acoustic radiation force of a focused vortex beam. Microbubbles approximately 100 µm in size can be displaced by a few microns in soft hydrogels with a typical shear modulus ranging from 10 to 100 Pa, consistent with a radiation force in the micronewton range. The applied force can be modelled using resolved measurements of the spatial properties of the incident pressure field. Using a simple elastic model for the bubble motion in the medium, the net displacement enables the experimental estimation of the applied force. We find a relatively good quantitative agreement between the radiation force model and our experimental discussion. Possible reasons for the discrepancies are discussed. Overall, the proposed approach presents real benefits compared to other conventional methods to assess the mechanical properties of soft and fragile materials, as it can be considered minimally intrusive, local, and well-adapted to probe thick and opaque-to-light materials in bulk.

Keywords: acoustic radiation force, acoustical tweezers, bubble dynamics, soft matter mechanics.

Introduction

Ashkin's pioneering studies of the radiation pressure of laser light has led to powerful tools to trap and manipulate small particles in optical tweezers [1] and have had a far-reaching impact in the the physical and life sciences [2, 3, 4]. The particle manipulation capabilities of optical tweezers rely on so-called single-beam gradient force traps in which a restoring radiation force is generated in all directions of space near the tight focus of a propagating beam. Recent research has allowed to extend the single-beam approach to propagating acoustic waves, for three-dimensional (3D) trapping in a liquid [5, 6], or acoustic levitation of particles against their own weight in air [7]. This work was mostly motivated by the considerably higher magnitude radiation forces, larger scale for the trapping working distance and trapping volume, and by the high penetrability in complex media [8]. These demonstrations

Journal of Physics: Conference Series **2904** (2024) 012008 doi:10.1088/1742-6596/2904/1/012008

have opened new perspectives for particle manipulation in the fields of fluid dynamics and biomedical applications of ultrasound [9, 10, 11, 12, 13].

Alongside the previously mentioned experimental developments in fluids, several studies have considered the radiation force arising on spherical objects of radius, a, that are embedded in elastic or visco-elastic materials [14, 15, 16, 17]. Because the induced displacement, u, of the object is determined by the mechanical properties of the embedding material, these studies represent novel quantitative and contactless approaches to assess the local mechanical properties of visco-elastic materials, and to probe subtle mechanical properties of complex fluids or soft solids [18, 19, 20], provided the applied radiation force is accurately modelled or experimentally calibrated. These studies have mainly exploited the pushing force generated by axisymmetric focused beams on objects located on-axis, and induce a displacement that is readily detected using conventional optical or acoustical tracking techniques due to its relatively large magnitude, $u/a \sim 1$, along the propagation direction. Therefore, though these techniques represent interesting alternatives to more standard mechanical characterisation tools such as traction assays or shear rheometry, they are still limited by their sensitivity to macroscopic properties of the propagating medium, and more research is required to fully extend their capability to probe local mechanical properties at the mesoscopic and microscopic scales.

In this paper we combine a radiation force model and experimental observations to explore the suitability of the highly-controlled force field obtained in vortex beams [6] to induce a small, three-dimensional, motion of a microbubble embedded in a soft hydrogel. Microbubbles are characterized by a strong acoustic contrast due to their high compressibility in soft hydrogels, and show the potential to probe local mechanical properties of complex soft materials, eventually at the microscale, under low amplitude and bio-compatible acoustic forcing.

Model for the acoustic radiation force and microbubble displacement

A theoretical framework for calculating the acoustic radiation force acting on an elastic or fluid sphere embedded in a soft elastic medium was recently proposed by Ilinskii *et al* [17]. In short, the incident, Φ_i, and scattered, Φ_s, acoustic fields, more precisely the compressional displacement potential fields, are first expanded in spherical coordinates using spherical Bessel functions and spherical harmonics. The total potential field $\Phi_t = \Phi_i + \Phi_s$ is then computed in the object's reference frame by applying appropriate boundary conditions on the object's surface, including partial conversion of the incident compressional field into shear waves (expanded in the same spherical basis). Under several assumption detailed in Ref.[17], particularly by omitting the contribution of scattered shear waves, the radiation stress acting on the object's surface and therefore the resulting radiation force, $\vec{F_A}$, can be computed without additional assumptions on the wavefront's spatial properties or particle size. To obtain the potential field Φ_i describing the incident field, we used beam shape coefficients (BSCs) originally derived to describe the expansion of vortex beams propagating in a fluid [5]. We here assume that such a description holds and allows to model the compressional component of the incident acoustic displacement field in the soft elastic medium. Moreover, the exact location of the particle in the incident field is managed through a combination of translations and rotations of the BSCs used to describe the initial configuration of a particle located on-axis at the beam focus (Ref. [21]). The final expression for $\vec{F_A}$ has been implemented using Python along with the rotation and translation transformations that are applied to the BSCs. More details on the implementation and a physical analysis of the expression for the force in the long-wavelength regime ($a/\lambda \ll 1$) will be provided in our future work under preparation.

Combining expressions for the radiation force with a model for the microbubble displacement allows to assess the mechanical properties of the soft hydrogel. Conversely, if the mechanical properties of the medium are known, for example by using shear rheometery, the model allows for an experimental calibration of the radiation force [18]. Such a model for the motion of a rigid sphere, or a deformable bubble, in incompressible elastic media has been proposed by Ilinskii *et al* [22]. Assuming a no-slip boundary condition on the object's surface (valid for a rigid sphere), the resistance force, F_R, applied

Journal of Physics: Conference Series **2904** (2024) 012008 doi:10.1088/1742-6596/2904/1/012008

as a result of the generated stresses in the medium reads:

$$\vec{F_R} = -6\pi G a \vec{u}, \tag{1}$$

where G is the quasi-static shear modulus of the medium. In our proposed experiments detailed bellow, the resisting force exactly offsets the applied acoustic radiation force, that is $\vec{F_A} = -\vec{F_R}$, when the microbubble reaches its final equilibrium position. While this expression is in principle not applicable to the motion of a bubble, it is still used in this preliminary work for sake of simplicity. The implications of this choice for the force calibration protocol will be discussed.

Acoustical tweezers, microbubble generation and displacement detection

A schematic of our experimental setup is depicted in Figure 1a. The acoustical tweezers apparatus consists in an array of 8 piezoelectric elements arranged as radial sectors on a concave surface (Imasonic, France), and have a central operation frequency f_0=2.25 MHz (wavelength $\lambda = 660\mu$m in water-like media). The array's aperture (F# =0.75) and the independent driving of each element allow to tightly focus a vortex beam at a focal distance $F = 40$ mm away from the active surface. The array is mounted at the bottom of a small water tank (30 x 20 x 15 cm^3) and emits the beam upwards. Vortex beams with a topological charge $\ell = 1$ have been shown to significantly increase the stiffness of the acoustic trap in all directions of space [5, 6]. We use a fibre-optic (Precision Acoustics, UK) mounted on a motorized 3D translation stage to measure the amplitude and phase of the pressure field on predefined points in space. The fiber's small diameter (100μm) and small active surface (10μm) is used to record the pressure and phase of the vortex beam in water or directly inserted in the hydrogels with a high resolution. Figure 1b and c show respectively the magnitude and phase maps that are obtained in a hydrogel by applying a Fourier transform to the pressure waveforms recorded in the focal plane (x, y) and analysed at the central frequency f_0.

A 3D-printed cuvette (PLA filament printer) was designed to contain the hydrogel and can be displaced in the water tank using the motorized stage. The bottom of the cuvette is sealed with an acoustical window made of a thin polyethylene film ($\sim$ 10μm thick) allowing the transmission of the ultrasonic beam directly from water into the hydrogel without distortion of its wavefront. The upper side of the cuvette is sealed with a microscope glass coverslip. The beam is focused several centimeters away from the coverslip to avoid perturbations of the propagating beam in the focal plane due to reflections. Finally, two small holes on the cuvette side allow to mount cylindrical plastic capillaries that guide either the fibre-optic, or the electrodes of the electrolysis system described bellow into the hydrogel.

The hydrogels are prepared by diluting a polymer (Carbopol ETD-2050) in distilled water at a low mass concentration (0.3 to 1%), following the protocol described in Ref. [23]. First, the water is heated to 50°C in a beaker hermetically sealed with parafilm and immersed in a thermal water-bath. Next, the right amount of polymer powder is progressively mixed while the solution is stirred with a magnetic agitator until its complete dissolution. When the mixture's temperature is back to ambient temperature, small crystals of sodium hydroxide are added until the solution's acidity is brought back to pH$\sim$ 7, as measured using pH paper. Once set, the hydrogels are sealed, centrifuged and stored for a few days. Their mechanical properties were characterised using standard oscillatory shear rheometry. Typical results for the storage shear modulus, G, as a function of frequency are shown in Fig. 1d. The yield stress of the hydrogels is also assessed with the rheometer and was found to be in the range of a few pascals for the different gel concentrations.

A microbubble is directly generated in the hydrogel using a custom made electrolysis setup. A first small copper electrode (144 μm in diameter) is inserted through the capillary and guided towards the center of the hydrogel cuvette. The second electrode is placed elsewhere in the water tank. Using a low amplitude electric field obtained with a 9-volt battery, microbubbles having a typically radius, $a \sim 50 - 100\mu$m are obtained by the successive coalescence of smaller bubbles at the electrode tip. The electrode is then quickly removed in order to ensure that the microbubble remains trapped near the

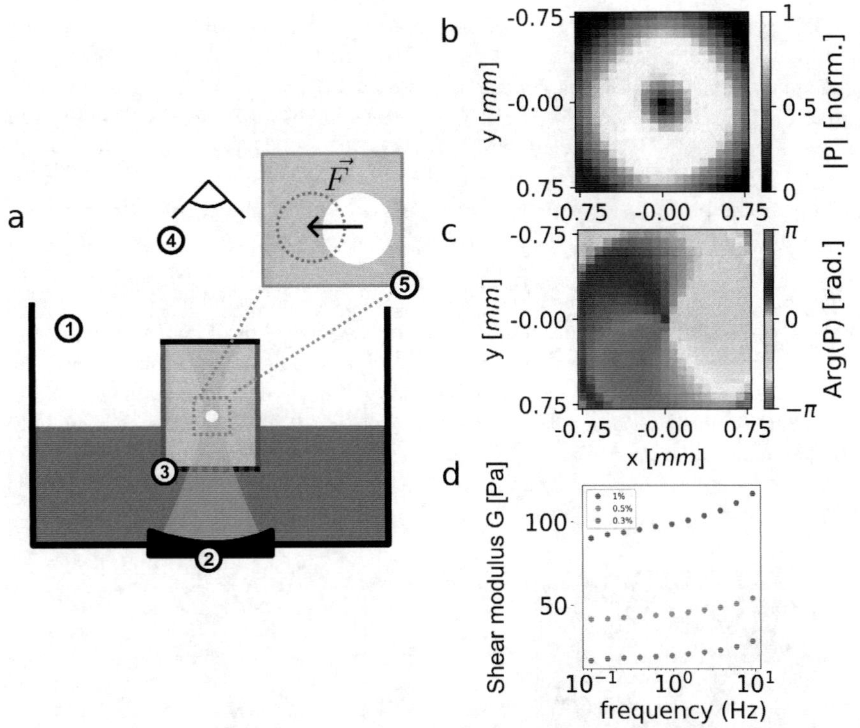

Figure 1. **(a)** Schematic of the experimental setup: (1) water tank, (2) 8-element piezoelectric transducer array, (3) mobile cuvette containing the hydrogel and a trapped microbubble, (4) optical system for microbubble motion detection (5). **(b)** Normalized magnitude $|P|$ of the incident vortex beam and phase map $Arg(P)$ **(c)**. **(d)** Shear modulus G of Carbopol hydrogels (0.3,0.5 and 1 % wight concentration) obtained with standard oscillatory shear rheometry with a typical error of 20%.

location it was generated, typically at the centre of the cuvette. It can further be manoeuvred back to the trapping plane by simply moving the cuvette. The optical focal plane of a custom made high-magnification optical system is brought to coincide with the acoustic trapping plane when the pressure field is scanned and the hydrophone tip is observed with a camera. Finally, the microbubble motion induced by the trapping beam can be detected with the same optical system. Its high magnification leads to a resolution of 1.1μm per pixel on the camera's sensor. An open-access image segmentation algorithm (Ilastik) is then used to detect the bubble in the RAW images. Fiji is sequentially used to detect the contour and centroid of the microbubble in the stack of segmented images. The bubble centroid can be obtained with a sub-pixel resolution around 0.4μm with our current lightning conditions.

Results and Discussion

The magnitude and phase of the focused vortex beam shown in Fig.1b and c indicate that there is no alteration to the wavefront as it passes through the cuvette's acoustical window. We also found no observable difference between the characteristics of the beam propagating in pure water or in the hydrogel, consistent with a liquid-like behavior of highly diluted hydrogels in the MHz regime [24]. Note that this observation is at the basis of our assumption that the incident compressional displacement field can be accurately modelled using a framework originally developed for beam propagation in fluid media [5]. The peak pressure amplitude measured on the vortex ring was set to $p_A = 90 \pm 16$ kPa. The phase varies from $-\pi$ to π around the vortex core (singularity) consistent with the topological charge $\ell = 1$ chosen to design the amplitude and phase law that is used to drive the 8 piezoelectric elements. Importantly, a limited variation of the phase along any radial line drawn from the vortex core outwards indicates that our field measurements were precisely performed at the geometrical focus of the concave array, F.

For microbubble trapping experiments, the same beam was emitted continuously for a few seconds. Before that, the microbubble is first positioned in the focal plane. Three independent images of the same bubble were obtained for each position respectively before ultrasound was applied, during acoustic forcing once the bubble reached its new equilibrium, and after its full relaxation once the emission was shut. Figures 2a and b show respectively a photograph of a microbubble being forced, and a schematic for the decomposition of its motion into radial and azimuthal displacement components that are observed in our experiments. By positioning the microbubble at different locations in the focal plane with a step size of 75μm, we were able to obtain high resolution maps of the displacement components. The results for a microbubble having an initial radius $a \sim 70\mu$m in a 0.3% hydrogel are shown in Figures 2c and d. Both components have a similar magnitude of a few microns. These minute displacements in the range ($u/a \sim 10^{-3} - 10^{-2}$) are typically 2 orders of magnitude smaller than in previously reported research [15, 25, 18].

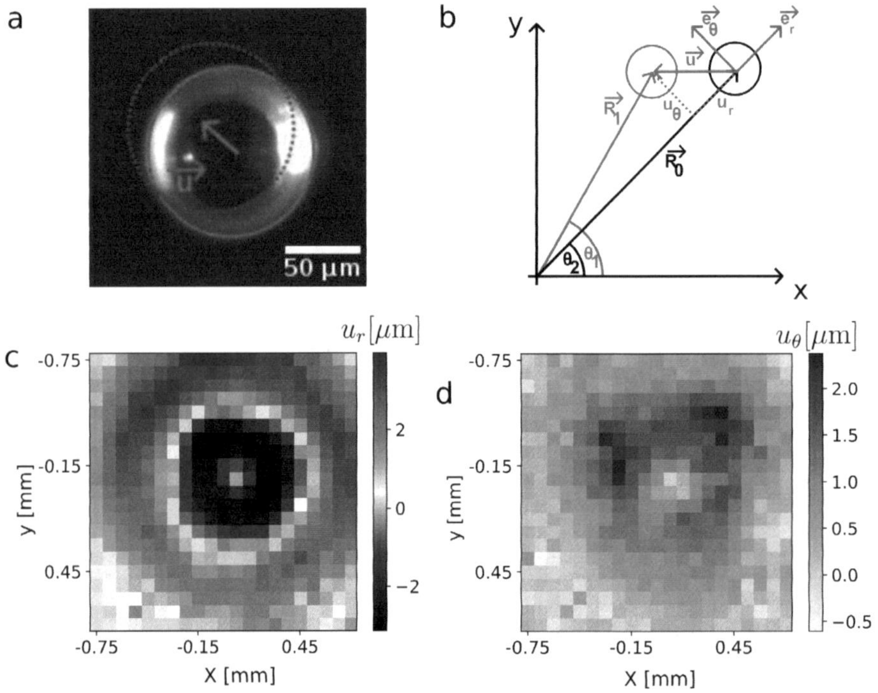

Figure 2. **(a)** Photograph of a microbubble (radius $a = 70\mu$m) in the trapping plane. **(b)** The microbubble displacement induced by the acoustic radiation force exerted by the vortex beam can be decomposed into radial (u_r) and azimuthal (u_θ) components. Spatial maps of the radial **(c)** and azimuthal **(d)** components obtained by displacing the hydrogel and microbubble in the trapping plane.

The "doughnut" symmetry of the vortex beam is clearly seen in the radial and azimuthal displacement components, and suggests that the mechanical properties of the hydrogel are isotropic. In the radial direction, the microbubble is pulled towards the vortex core up to a radial distance, $r \sim 450\mu$m away from the vortex core that coincides with the region of maximum pressure magnitude.

Journal of Physics: Conference Series **2904** (2024) 012008 doi:10.1088/1742-6596/2904/1/012008

Beyond this position, the microbubble is instead pushed away from the core. This behavior is consistent with a description of the radial radiation force component as a "gradient" force. In this description, the force is directly proportional to the local gradient of the potential energy density [6, 26] for a microbubble undergoing monopolar oscillations [11]. Because the microbubbles are here driven well above their natural monopolar resonance frequency (around 50 kHz in water for the typical bubble size in our experiments), they are forced towards local "nodal" regions where the pressure magnitude vanishes, as previously observed for bubbles in liquids [11]. However, because the microbubbles have a non-negligible size ($a/\lambda \sim 0.1$), they are also sensitive to a non-conservative, scattering force component, leading to the non-zero azimuthal displacement (Fig.2d). In contrast with the radial component, it is maximum on the vortex ring, consistent with a scattering force component proportional to the local azimuthal energy flux generated by acoustic fields having a non-zero orbital angular momentum [21][24]. Nevertheless, the microbubbles remained small enough, and we did not detect any "size effects" on the radiation force that are usually observed for higher order resonances that appear in the so-called Mie scattering regime ($a/\lambda > 1$), as recently observed in the context of acoustic levitation [27]. Conversely, the azimuthal scattering force component is expected to decrease much faster compared to the gradient in the long-wavelength regime ($a/\lambda \ll 1$).

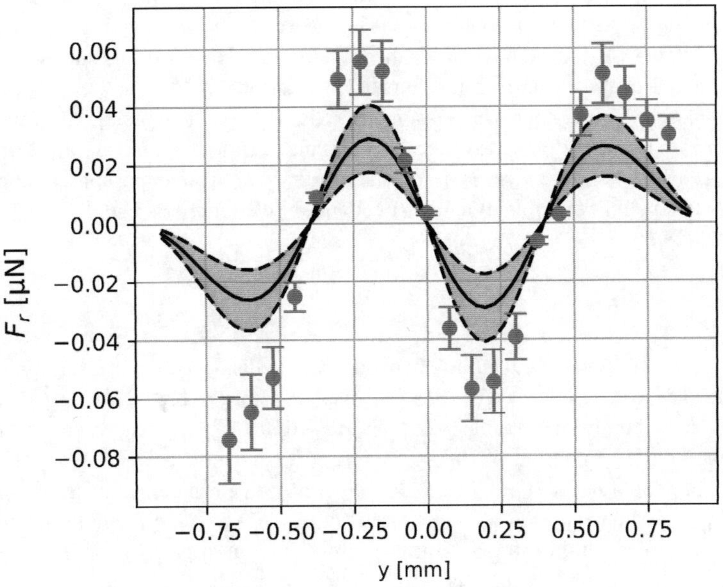

Figure 3. Radial component of the acoustic radiation force exerted on a microbubble of radius $a = 70\mu$m. The plain curve is obtained with a radiation force calculation using for the peak pressure $P_A = 90$ kPa. This curve is bounded by two other calculations that apply a $\pm 14\%$ uncertainty to the pressure measurement. The data points are obtained from the radial displacement component shown in Fig.2c and using equation (1). The error bars are obtained by applying a 20% uncertainty to the measured value of the hydrogel's shear modulus.

Finally, we aim at exploiting the observed microbubble motion to test the accuracy of the acoustic radiation force and bubble displacement models. The spatial properties of the incident pressure field and its magnitude that have been measured are first accurately reproduced with the propagation model described in the first section (data not shown). The theoretical curve obtained for the radial force component, F_r, is shown in figure 3 for a microbubble mean radius, $a = 70\mu$m. The continuous line is obtained for $p_A = 90$ kPa and is bounded by two other calculations obtained by applying a 14% uncertainty to the peak pressure magnitude due to the instrumental error. Using equation (1) with a prior knowledge of the shear modulus, G, of the hydrogel taken at the low frequency limit (Fig.1d), the experimental force component can be estimated from the radial component of the microbubble displacement (Fig.2c). The error bars represent the 20% uncertainty we applied to the measured value of the shear modulus, G. Note that the error in determining the bubble size with our optical system is in comparison negligible. Overall, our results are in a relatively good quantitative agreement with the radiation force model that we used. The asymmetry observed for the experimental force can be ascribed to a slow dissolution of the microbubble as it was scanned through the trapping plane. Currently, either the calculation underestimates the actual radiation force or the mechanical model for the microbubble displacement in the hydrogel is overestimating it. In the first case, it may be possible that the conversion of compressional waves into shear waves upon scattering on the bubble has an actual effect on the radiation force magnitude that has been neglected in the model [17], though the conversion should be small for bubbles mainly responding to the incident field with a monopolar oscillation. Additionally, we currently do not control the exact gas content of the bubbles, though we estimate this should have a negligible effect on its actual compressibility, it could still reflect in the estimation of the force. Regarding the mechanical model estimating the displacement, one important assumption is that the hydrogel behaves an elastic medium. While in most cases we were able to check that the bubble would indeed relax back to its initial position, for the smallest bubbles, the stress exerted on the medium during its motion would in some cases exceed the hydrogel's yield stress [19, 20], leading to an irreversible plastic deformation of the hydrogel. These observations suggest that a more complex model for the hydrogel mechanics could improve our results. Additionally, the expression for the displacement in equation (1) considers the motion of a rigid object. The possible deformation of the bubble interface during its motion is expected to modify the magnitude of the stresses generated in the hydrogel [22]. More experiments are underway in our laboratory to reduce experimental uncertainties, and advanced models for the microbubble dynamics in complex soft materials could also be considered in our future work [20, 28].

Conclusion

We have shown that microbubbles embedded in a hydrogel can be displaced by the radiation force field generated in an ultrasonic vortex beam. We detected displacements of a few microns for bubbles a hundred microns in size in highly deformable gels ($G \sim 10 - 100$ Pa). By accurately controlling the position of the bubble and the direction of the applied force in acoustical tweezers, our experiments show the potential to develop a novel approach to spatially map the mechanical properties of unknown complex materials. The method relies on a contactless driving force and is minimally-invasive. Using smaller bubbles, this approach could result relevant to probe local, microscale, and complex mechanical properties of a variety of soft materials, including their inhomogeneous, anisotropic and active nature.

Aknowledgements

We would like to thank T. Salez and Y. Amarouchene for their assistance in characterizing the mechanical properties of the hydrogels. We also express our gratitude to P. Lidon, G. Ovarlez and B. Saint-Michel for fruitful discussions on the rheology of Carbopol hydrogels.

References

[1] Ashkin A, J M Dziedzic, Dziedzic J M, J M Dziedzic, J M Dziedzic, Bjorkholm J E and Chu S 1986 *Optics Letters* **11**
[2] Grier D G 1997 *Current opinion in colloid & interface science* **2** 264–270
[3] Yao A, Tassieri M, Padgett M and Cooper J 2009 *Lab on a Chip* **9** 2568–2575

[4] 1994 *Annual review of biophysics and biomolecular structure* **23** 247–285
[5] Baresch D, Thomas J L and Marchiano R 2013 *Journal of Applied Physics* **113** 184901
[6] Baresch D, Thomas J L and Marchiano R 2016 *Physical Review Letters* **116** 024301
[7] Marzo A, Seah S A, Drinkwater B W, Sahoo D R, Long B and Subramanian S 2015 *Nature communications* **6** 8661
[8] Thomas J L, Marchiano R and Baresch D 2017 *Journal of Quantitative Spectroscopy and Radiative Transfer* **195** 55–65
[9] Melde K, Mark A G, Qiu T and Fischer P 2016 *Nature* **537** 518–522
[10] Baudoin M, Gerbedoen J C, Riaud A, Matar O B, Smagin N and Thomas J L 2019 *Science Advances* **5** eaav1967
[11] Baresch D and Garbin V 2020 *Proceedings of the National Academy of Sciences* **117** 15490–15496
[12] Ghanem M A, Maxwell A D, Wang Y N, Cunitz B W, Khokhlova V A, Sapozhnikov O A and Bailey M R 2020 *Proceedings of the National Academy of Sciences* **117** 16848–16855
[13] Baudoin M, Thomas J L, Sahely R A, Gerbedoen J C, Gong Z, Sivery A, Matar O B, Smagin N, Favreau P and Vlandas A 2020 *Nature Communications* **11** 4244
[14] Aglyamov S R, Karpiouk A B, Ilinskii Y A, Zabolotskaya E A and Emelianov S Y 2007 *The Journal of the Acoustical Society of America* **122** 1927–1936
[15] Orescanin M, Toohey K S and Insana M F 2009 *The Journal of the Acoustical Society of America* **125** 2928–2936
[16] Lopes J, Leão-Neto J and Silva G 2017 *The Journal of the Acoustical Society of America* **142** 2866–2872
[17] Ilinskii Y A, Zabolotskaya E A, Treweek B C and Hamilton M F 2018 *The Journal of the Acoustical Society of America* **144** 568–576
[18] Lidon P, Villa L, Taberlet N and Manneville S 2017 *Applied Physics Letters* **110** 044103 arXiv:1611.08633 [cond-mat, physics:physics]
[19] Lidon P, Villa L and Manneville S 2019 *Soft matter* **15** 2688–2702
[20] Saint-Michel B and Garbin V 2020 *Soft Matter* **16** 10405–10418
[21] Baresch D, Thomas J L and Marchiano R 2013 *The Journal of the Acoustical Society of America* **133** 25–36
[22] Ilinskii Y A, Meegan G D, Zabolotskaya E A and Emelianov S Y 2005 *The Journal of the Acoustical Society of America* **117** 2338–2346
[23] Geraud B, Bocquet L and Barentin C 2013 *The European Physical Journal E* **36** 1–13
[24] Alhaïtz L, Brunet T, Aristégui C, Poncelet O and Baresch D 2023 *Physical Review Letters* **131** 114001
[25] Yoon S, Aglyamov S R, Karpiouk A B, Kim S and Emelianov S Y 2011 *The Journal of the Acoustical Society of America* **130** 2241–2248
[26] Gor'kov L P 2014 On the forces acting on a small particle in an acoustical field in an ideal fluid *Selected Papers of Lev P. Gor'kov* pp 315–317
[27] Ospina J F P, Contreras V, Estrada-Morales J, Baresch D, Ealo J L and Volke-Sepúlveda K 2022 *Physical Review Applied* **18** 034026
[28] Oratis A T, Dijs K, Lajoinie G, Versluis M and Snoeijer J H 2024 *The Journal of the Acoustical Society of America* **155** 1593–1605

AUTHOR INDEX